GEOTHERMAL RESOURCES

GEOTHERMAL RESOURCES

ROBERT BOWEN
Ph.D., B.Sc.

Formerly Consulting Geologist
Tehran Berkeley Civil and Environmental Engineers
Tehran, Iran

APPLIED SCIENCE PUBLISHERS LTD
LONDON

APPLIED SCIENCE PUBLISHERS LTD
RIPPLE ROAD, BARKING, ESSEX, ENGLAND

British Library Cataloguing in Publication Data

Bowen, Robert
 Geothermal resources.
 1. Geothermal engineering
 I. Title
 621.4   TJ280.7

 ISBN 0-85334-846-4

WITH 21 TABLES AND 49 ILLUSTRATIONS

Printed in Great Britain by Galliard (Printers) Ltd, Great Yarmouth

Preface

Until recent years, it has been taken for granted that unlimited quantities of low-cost energy would always be available, but it is becoming increasingly clear that in order just to maintain existing standards of living in the developed countries, fossil fuels must be supplemented from other sources so that events such as those leading to the Arab Oil Embargo of the winter of 1974 will not again easily threaten to bring our technological civilisation to a standstill. The USA is particularly vulnerable because it consumes one-third of the world's energy production even though it has only 6% of the population of the Earth. The problem becomes more acute when it is realised that planetary consumption of energy derived from fossil fuels such as coal and oil is increasing at a rate of approximately 2% annually.

Alternative sources include solar energy and geothermal energy and it is with this latter that the present book is concerned. Italy has been generating electricity from natural steam sources since 1904, but The Geysers steam plant in California constitutes the largest geothermal installation in the world at present and it produced almost 1000 MW in 1976. The USA is, in fact, the largest producer of electricity from geothermal energy on Earth and a report by the National Science Foundation forecast that the country could be generating 137 000 MW of electrical power from geothermal sources by 1985. As this implies, geothermal resources are known to exist in many other areas of the United States than at the Big Sulphur Creek in California where The Geysers are located. Several other countries, notably New Zealand, Mexico, Japan, Iceland and the USSR, are all directing increased attention to exploitation of their own geothermal resources. Additionally, the potential of these is not restricted to generating electrical power because mineral extraction, fresh-water production and space and process heating also constitute possible benefits from their application under appropriate circumstances.

The source of heat lies within the Earth and considerations of plate tectonics, a recent revolution in geology, are involved and discussed

v

initially. Understanding these is important in comprehending why geothermal resources occur where they do. Geothermal systems are described and classified, the presently known significant ones being noted. Exploration methods which have led to the discovery of new geothermal fields deserve treatment in some depth and in attempting this, the writer has indicated some useful recent approaches such as the use of satellites, radiogenic noble gases in geothermal tracing and microwave radiometry.

Subsequently, hydrothermal convection systems are examined and then the artificial stimulation of geothermal systems by the use of conventional and nuclear explosives. Later chapters are devoted to The Geysers and the geothermal resources of New Zealand, including Wairakei which first supplied electricity from a wet-steam-field resource in November 1958. Iceland is the main subject of Chapter 8 which deals with space and process heating. Succeeding chapters cover geothermal resources of the world other than the major ones previously described, the relationship of these to water (with desalination possibilities), recent developments in geothermics in Italy, environmental effects, current researches and future prospects. A glossary of some of the more important technical, geological and geothermal terms used in the book appears at the end. There are a number of tables, figures and plates, acknowledged where necessary.

Although growing apace, geothermal literature is somewhat fragmentary at the moment. Rather than append a single alphabetical bibliography listing a mixture of relevant topics terminally, the writer has grouped the most important references after each chapter and they are numbered as they appear within.

The future holds much promise for geothermal energy which, it has been suggested, may contribute as much as 10% of total western USA electrical power consumption within the next few years. There are many advantages to utilising it where it can be found. It is not a short-lived resource, contributes less to pollution than other sources of energy, is economically competitive with conventional sources of electricity, can be stimulated artificially and topped up (thus being, to an extent, renewable unlike fossil fuels), does not cause subsidence and seismic problems which mining activities do, has low manning costs and is indigenous, hence not subject to foreign interference. In the years ahead, technological advances in deep drilling may well enable a wider development of the geothermal resource to take place, one in which Man may penetrate further into the planetary crust than the 8000 ft achieved by some wells at The Geysers.

ROBERT BOWEN

Contents

CHAPTER 1

Geology and the Source of Heat

Since the nineteenth century, Man has become increasingly dependent upon fossil fuels to maintain and improve his standard of living, and the economic ease of obtaining these to date has led to a blindness as regards the future. Not so many years ahead, oil and coal, non-renewable resources in human terms, will have run out and in any event, as the Arab embargo of the former in 1974 demonstrated, political factors can deprive us of them at any time. Also, the often deleterious environmental effects of our existing energy sources have been ever more apparent in recent years. Oil spills, strip mining, the emission of sulphur and accumulation of solid wastes have emphasised the need for clean energy. Even nuclear power, no longer thought of as the solution to the problem, is suspect as regards pollution.

Examining the petroleum question, over the last decade the American rate of consumption has been increasing at an annual rate of 4·2 % and the dependence of the USA on imported oil reached more than a third by 1974. Add to this the fact that no new field has been discovered in the original 48 states containing as much as a thousand-million barrels reserve and it is clear that the ever-widening gap between production and consumption in America is a logical consequence and one with serious implications in view of the yearly growth rate in consumption noted above. It has been proposed that a conservation growth rate of 2·1 % per year would remedy matters, but it would not. This is because more oil would still be used than the quantity available and the only result could well be a slowdown in industry.

Obviously, alternative means of energy production must be developed if living standards are to be maintained, let alone improved, and these ought to be of such a nature that ill-effects on the environment can be minimised. Figure 1.1 shows the energy balance for the United States in 1970 and forecasts indicate that fossil fuels, dominant then, will continue to be so at least until 1985. Thereafter, one alternative may well make a significant

1

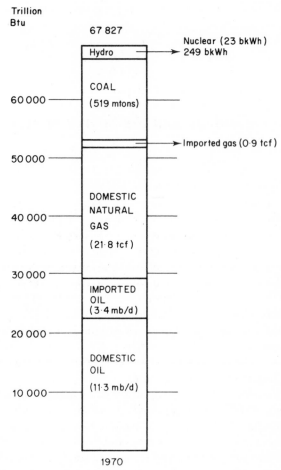

FIG. 1.1. U.S. energy balance, 1970, by types and origins of resources. mtons = millions of tons; tcf = trillions (10^{12}) of cubic feet; bkWh = billions (10^9) of kilowatt-hours; mb/d = millions of barrels per day. Note the absence of geothermal power and power from natural gas produced by coal gasification.

contribution and that is geothermal energy. Others, such as solar energy, oil from shale and hydroelectric power will also become important. However, by the middle 1980s, all the 'new technology' sources together are not expected to furnish more than 3% of American energy needs.[1]

Nevertheless, these new sources will increase in significance simply because proven planetary petroleum reserves will probably be depleted within 70 years.[2] This is a good argument for diverting financial resources to suitable exploration activities.

The Earth's natural heat is geothermal energy. Heat is, in fact, conducted out from the interior of the surface of the planet at an average rate of approximately $1 \cdot 5 \, \mu \text{cal/cm}^2$ s and, over a time interval of a year, this flux to the entire surface exceeds 10^{20} cal. D. E. White and D. L. Williams estimated that the heat stored in rock beneath the USA to a depth of 10 km is of the order of 8×10^{24} cal.[3] Of course, a lot of this heat is not usable. Practically, heat must be concentrated in geothermal reservoirs where it is built up and stored through geological processes in order for it to be exploitable. It is interesting to observe, however, that in the upper 10 km (where the temperature exceeds 100 °C) the total stored geothermal energy exceeds, by orders of magnitude, all thermal energy available in all nuclear and fossil-fuel sources.

In order to understand the origin of the heat, heat-flow measurements, both on the continents and under the seas, have been effected all over the world in recent years and these, coupled with the relatively new interpretation of the crust in terms of plate tectonics, have facilitated this.

The latter, discussed by the writer in an earlier paper,[4] must now be examined in some detail.

1.1. A REVOLUTION IN GEOLOGY

That continents move is quite an old idea in human thought and it was Francis Bacon in *Novum Organum* who first pointed out the conformity of outline between the Atlantic coast of Africa and the Pacific coast of South America. Early in the last century, Alexander von Humboldt noted the congruence of the west African and eastern South American coasts and thought that the Atlantic represented a vast valley excavated by the sea. However, the first appearance of the concept of an actual splitting and drifting apart of these two continents was in *La Création et ses Mystères Dévoiles* by Antonio Snider-Pellegrini. He believed that when the Earth cooled from an initial molten state, continents formed only on one side causing instability later relieved during Noah's Flood by a catastrophic fracturing and separating of the Americas from the Old World. Catastrophism, cited to explain the origin of the Moon from the Pacific by G. H. Darwin in 1879, continued to intrude into ideas of continental drift into this century. A new departure came from the work of Alfred Wegener. At the turn of the century, the general opinion was that the Earth was still in the process of solidifying and contracting from the initial state alluded to by Snider-Pellegrini and that, as a result, lighter rocks rich in silicates of alumina together with potash and soda (sial) such as granitic-type igneous and metamorphics with associated sediments had migrated to the crustal surface (Fig. 1.2). Underlying them were denser rocks similar to basalt or gabbro which constitute sima. Mountain ranges were believed to form by contraction, cf. the crinkling of the skin on a drying (shrinking) apple.

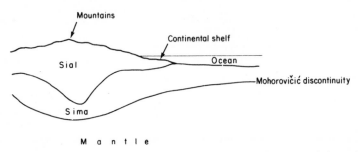

FIG. 1.2.   Cross-section of the crust of the Earth showing lighter sial overlying denser sima and comprising the continent, the sima underlying both that structural unit and the oceanic bottom. The mantle underlies the crust.

Arching pressures caused subsidence of sectors of the planetary surface to form the oceans, and the remaining regions, the continents, continued emerging as great blocks. Trans-oceanic land bridges were imagined as formerly existing in order to explain similarities between many fossils, both animal and plant, now found on different continents. From the stratigraphic record, rises and falls in sea level could be inferred, Eduard Suess's eustatic changes, and accompanied marine transgressions and regressions over the continents. Regressions were attributed to oceanic basin subsidence, and transgressions to partial infilling of these basins as a consequence of continentally derived sedimentation. This model became quite deeply entrenched until Wegener demolished it. Lauge Koch, one of his associates in Greenland expeditions, said that the great German meteorologist first obtained his idea of continental drift from observing the splitting and separating of ice slabs in the sea. Wegener himself stated that he thought of it by noting the congruence of coastlines on either side of the Atlantic Ocean. Afterwards, actually in 1911, he read that palaeontological evidence existed for a former land bridge between Brazil and Africa. This led to publication of several papers and his famous book translated in 1924 *The Origin of Continents and Oceans*. At this stage, many arguments were

FIG. 1.3. A reconstruction of Pangaea. This is the hypothetical super super-continent which is thought to have commenced breaking up about 200 million years ago and later produced the supercontinents Laurasia (in the northern hemisphere) and Gondwanaland (in the southern hemisphere).

possible against the contracting Earth hypothesis. The enormous lateral thrusting of rock slices (nappes) in the Alps led to estimates of Tertiary† contractions which appeared to be excessive. Anyhow, why were fold wrinkles so irregularly distributed over the planetary surface, localised in narrow belts? Continentally located marine sedimentary rocks are nearly always shallow-water in type and could not have been formed in oceanic deeps. Thus, Lyell's idea that continents and oceans are interchangeable could no longer be accepted. Gravity data indicated that sima underlies the oceans and sial the continents. This and the concept of isostasy, the tendency of the crust to maintain a state of near equilibrium, rendered impossible the subsidence of continental regions into oceanic deeps. Wegener postulated that, in the Mesozoic, a super supercontinent, Pangaea (Fig. 1.3), rifted and subsequently, indeed up until the present, its fragments separated. South America and Africa drifted apart in the

† The chief divisions of geological time are shown in Appendix 1.

Cretaceous and so did North America and Europe although they kept contact in the north up to Quaternary times. He suggested that the western Cordilleran ranges had been formed by compression at the leading edges during the westward drift of the Americas, that, as India moved northwards, land in its path had been crumpled up to form the Himalayas, and that New Guinea and Australia had separated from Antarctica in the Eocene and moved northwards also, entering the Indonesian archipelago in late Tertiary times. Mostly, he concentrated on the two sides of the Atlantic and this is certainly a fruitful area. The Cape fold belt in South Africa seems to continue in the range of Buenos Aires province in Argentina. The Upper Palaeozoic–Lower Mesozoic non-marine series of the Karroo System in South Africa parallels in many respects the Santa Catharina System in Brazil. The antique gneiss plateau of Africa resembles that of Brazil in various kinds of igneous rocks and kimberlite. These geological phenomena exist also in the North Atlantic and can be supplemented by palaeontological and biological data as well. Palaeoclimatic arguments may be cited too. The most significant geological evidence relevant to the matter is given by glacial boulder beds (tillites) lying on striated pavements of resistant rock. They represent traces of formerly existing ice sheets. Coal beds are also useful, indicating former humid conditions. As regards the former, Carboniferous and Permian glacial deposits have been found in South America, South Africa, India and Australia, components of a southern hemisphere division (one of two) of Pangaea, namely Gondwanaland. Wegener also introduced the idea of polar wandering.

Curiously, Wegener's ideas did not find much general acceptance or even fame before the second world war and indeed some of his critics were very hostile. His jigsaw fit of the Atlantic continents was challenged on the grounds that it was inaccurate which, given epeirogenic (vertical tectonic) movements frequently occurring along coastlines as they do, might well have been anticipated. Charles Schuchert preferred land bridges to explain the faunal and floral similarities between continents. Wegener's opinion that the latter tend to move towards the equator, the famous pohlflucht, was particularly attacked by the Cambridge geophysicist Harold Jeffreys who believed that the planet has too much strength to be deformed by pohlflucht forces. In other words, indefinite deformation of the Earth by small forces acting over long geological time intervals was considered impossible as was polar wandering. He favoured a contracting Earth model based on condensation as a liquid from a gas cloud, subsequent convective cooling occurring after gravitational separation had quickly created both crust and core. Another hypothesis was presented in T. C. Chamberlain's planetesimal ('cold accretion') model, i.e. compaction under gravity coupled with a rise of temperature of small meteoritic bodies, an idea which was modified and represented in the late 1950s by Harold C. Urey. Despite the opposition of many, however, Wegener was supported by some noted authorities such as R. A. Daly, the Harvard geologist, and E. B. Bailey, the British structural geologist. By far the most doughty pre-second world war support came from Arthur Holmes and Alexander du Toit, the former a British and the latter a South African geologist. The publication of du Toit's

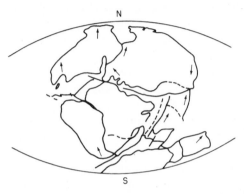

FIG. 1.4.   Earth 135 million years before the present. Laurasia to the north, Gondwanaland to the south, Tethys between.

Our Wandering Continents in 1937 pointed out the capacity of the continental drift idea to explain coherently a wide spectrum of geological phenomena and amassed additional supporting evidence. du Toit believed that Pangaea separated into a northern and a southern part, the latter Gondwanaland alluded to above and the former Laurasia (Fig. 1.4). These were separated, at least since the Upper Palaeozoic, by an extensive seaway called Tethys which did not break up until the Tertiary when Africa and India were driving northwards in the direction of Eurasia. After the end of the second world war, progress accelerated because more information became available regarding the oceanic floors. Another then new research field was very significant and that is palaeomagnetism.

1.1.1. Palaeomagnetism
Palaeomagnetism is sometimes referred to as rock magnetism and it relates to the magnetism of rocks. In rocks, magnetisation is actually a fossil permanent magnetism, i.e. a natural remanent magnetism (nrm) which acts

like a fossil compass enabling the direction of the ancient (palaeomagnetic) field to be determined. The assumption employed is that the mean geomagnetic field is that of an axial dipole situated at the centre of the planet. If a sensitive magnetometer is used to find out the angles of declination and inclination, D and I, for oriented rock samples, the angle of latitude is obtained from

$$\tan I = 2 \tan L$$

(L = angle of latitude). Not many rock types can be employed, but amongst those that can are basaltic lavas which are rather rich in iron and acquire their magnetisation from the geomagnetic field as they cool down through the Curie points of their iron oxide–titanium oxide minerals on crystallisation after eruption at the surface of the Earth. A few sedimentary rocks such as red sandstones also contain sufficient iron oxide to possess measurable magnetisation. It must be noted that magnetisation may also be acquired *after* deposition, i.e. during diagenesis. In practice, determination of D and I is effected after orientation of the rock sample into the position it would have occupied when formed so that any effects due to subsequent tectonic dip disturbance can be eliminated.

The axial dipole nature of the geomagnetic field accords with the theory of Sir Edward Bullard and W. M. Elsasser that the fluid and electrically conducting outer core of the planet is a body acting as a self-exciting dynamo. At present, the geographic rotational axis of the Earth is at an angle of about $11\frac{1}{2}°$ with the geomagnetic axis, but this was not always so. Over long periods, namely millennia, they coincide. As early as the 1950s, a Cambridge group consisting of S. K. Runcorn, K. M. Creer and E. Irving had demonstrated that, prior to the Upper Tertiary, there was a steady change with time in the position of the North Pole (Fig. 1.5). Located near Hawaii in the Precambrian, it slowly migrated northwestwards to between Kamchatka and Japan by the termination of the Palaeozoic and thereafter it moved on through eastern Siberia to its current position. Some people called it the drunken pole! Differences compared with now were found to be not random but rather systematic and it was inferred that, since the angle of polar difference increases with time, migration either of the poles or of the continents had taken place. Palaeomagnetic estimates that Europe lay between 20 °N and 20 °S in the Upper Palaeozoic are in good agreement with geological evidence of warm, humid climates, for instance coal beds and coraliferous limestones in the Carboniferous and evaporites and arid sandstones in the Permian. Later, in the Jurassic, palaeomagnetically determined latitudes for Europe up to 20 ° less north than now accord with palaeontological evidence again of warm, humid conditions conducive to the presence of reef-building corals and giant reptilia.

The preceding investigations of the Cambridge group related to European observations, and later they turned their attention to North American rocks from which a similar pole-wandering path was determined, but with a displacement of about 30° of longitude to the west! This was apparent in the Precambrian and Palaeozoic, but disappeared after the

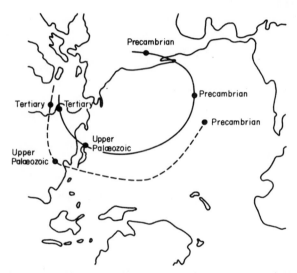

FIG. 1.5.   Polar wandering curves from Precambrian to today in Europe (solid line) and North America (broken line). Pole positions at stated times indicated by solid circles.

Trias. This anomaly is obviated if North America is brought adjacent to Europe by closure of the Atlantic, and Runcorn thus came to the view that separation of the two continents must have taken place between Triassic times and the present. He also became an apostle of lateral migration of continents. The same thing was found to be the case in the southern hemisphere and divergences of polar wandering paths for the various continents concerned could be eliminated if these were assembled as Gondwanaland.

Needless to say, many new adherents to continental drift resulted from all this. Trouble did arise occasionally, however, as a consequence of the fact that *secondary* magnetisation may occur, i.e. a rock may acquire a second dosage of magnetisation long after its formation. It was discovered that this can be removed either by magnetic cleaning using alternating magnetic fields or by thermal cleaning (heat treatment).

One of the valuable discoveries of palaeomagnetism is that geomagnetic reversals of polarity of short duration take place. Another relates to the ocean floors and depends upon post-war advances in both geological and geophysical oceanography.

1.1.2. Oceans

Oceans comprise three topographic provinces, namely continental margins (i.e. continental shelf, slope and, locally, deep trenches), ocean-basin floors (i.e. abyssal floors, oceanic rises and sea-mounts) and mid-oceanic ridges. Roughly, each accounts for about one-third of the total area. Naturally, they are huge because the oceans make up almost three-quarters of the planetary surface.

Continental shelves and margins have shapes determined in part by faults and in part by deposits of shallow-water sediments. The abyssal plains seem to be mainly smooth, sediment-filled basins. By far the most spectacular feature is the mid-oceanic ridge system with rugged topography reminiscent of high mountain ranges on land. Taken together, the ridges comprise a continuous swell traceable for 35 000 miles through the oceans according to Bruce Heezen. In some areas, they are very wide, up to 1000 km, for instance, in the East Pacific Rise. Their axial region is correlated with a shallow-focus earthquake zone. In some parts, its relief suggests nothing so much as a central rift valley.

Crustal structure in general has been extensively studied by techniques of seismic reflection and refraction. These depend on the fact that interfaces of rock strata are quite frequently very good reflectors and high sonic frequencies are employed in upper layers. Deeper down, however, these cannot penetrate and here low sonic frequencies are used, optimally together with gravity determinations.

The base of the crust is known as the Mohorovičić discontinuity (the well-known Moho) and below this velocities of seismic wave propagation increase suddenly to more than 8 km/s. The depth to this base is, of course, less under the oceans than under the continents and it *averages* 35 km.

Three crustal layers are usually distinguished (Fig. 1.6). These are as follows:

1. Layer 1 at the top is approximately 1 km thick and has a transmission velocity of some 2 km/s. It is thought to comprise sedimentary rocks.

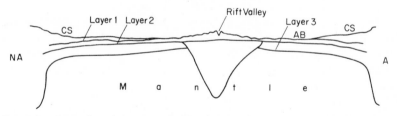

FIG. 1.6. Crustal section across the North Atlantic. This shows the Mid-Atlantic Ridge and Rift Valley. NA = North America; A = Africa; CS = continental slope and shelf; AB = abyss.

2. Layer 2 is approximately 1·7 km thick and has a transmission velocity 5·1 km/s. Considered to be either consolidated dense sediment or a modified version of layer 3.

3. Layer 3 averages 5 km in thickness and is usually believed to consist of either basalt or gabbro, i.e. basic igneous material.

Under the crust is a layer known as the mantle (Fig. 1.7) and this is thought to be composed of the ultrabasic rock peridotite under the oceans. Mostly, this is made up of olivine and pyroxene. Its transmission velocity is, as noted above, over 8 km/s. Under continents, it is believed that another rock, eclogite (again basic containing garnet and pyroxene and sometimes amphibole with accessory sphene, magnetite and zoisite), may occur. It is interesting that under the mid-oceanic ridges, a lens-shaped zone with slightly lower transmission velocities (between 7 and 8 km/s) is found to

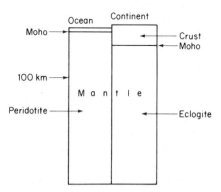

FIG. 1.7. Crustal thickness variations under continents and oceans showing variation in depth of Moho.

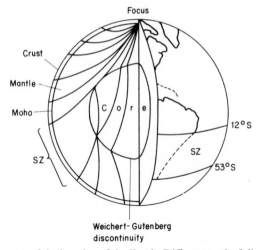

F IG. 1.8.   Structure of the interior of the Earth. Different paths followed by energy radiated by an earthquake near the North Pole are shown and, by analysis of these, the interior can be divided into a core and a mantle overlaid by a thin crust. Refraction in the core causes a shadow zone (SZ) in which the earthquake cannot be detected by waves channelled along the surface.

exist. It may be added that about 2000 km below the mantle surface is another low-velocity layer, the Gutenberg discontinuity (sometimes called the Weichert–Gutenberg discontinuity), separating mantle from core (Fig. 1.8).

Heat-flow measurements have also been made and the rate of flow of heat through the oceanic floor depends upon the product of temperature gradient and thermal conductivity. Temperature gradient is measured by inserting a probe containing a recorder with a temperature-sensitive element into the ocean floor. Thermal conductivity is measured by collecting samples of sediment by coring. The same approach is applied to measurement of heat flow on land, of course.

Surprisingly, average values of heat flow through the oceanic floors seem almost the same as those determined on continents. Since continental rocks have much higher radioactivity, this was most unexpected and the explanation may well be that sub-oceanic mantle may differ from sub-continental mantle and contain more radioactivity at depth.

Heat flow under the axis of the mid-oceanic ridge system was found to be higher than elsewhere in the oceans and this might well be expected in view of the evidence of volcanicity, e.g. in Iceland which straddles the Mid-Atlantic Ridge.

Turning now to measurements of the total magnetic field at sea, the normal technique is to tow magnetometers with gyrostabilised platforms behind ships or planes at a distance far enough to avoid magnetic disturbances due to the engines or hull. Mostly, the measurements are made on magnetite-bearing basalts on the oceanic floor. Magnetic anomalies are recorded graphically and they represent milligauss deviations from *mean* values of total magnetic intensity measured in the direction of the geomagnetic field.

The writer can well remember the excitement felt while he was at the Scripps Institute of Oceanography, La Jolla, California, in the late 1950s when the method was being used to obtain very significant results. A persistent north–south alignment of the anomalies giving, on east–west traverses, a characteristic oscillatory pattern of positive and negative values about the mean was found and the anomalies were thought to be delineatory of elongate bodies of magnetite-bearing rock (thought to be basalt) aligned north–south. Their patterns of magnetisation were in distinct contrast with their neighbours to the west and east.

The eastern Pacific is traversed by several major topographic features which have an east–west alignment, e.g. the Mendocino Escarpment off California, and they suggest fault control. Comparable structures displace the axial ridge in the Atlantic. It has been found that the unique oscillatory pattern of given anomaly traces can be very accurately matched across these possible fracture zones, but only as a result of invoking large-scale lateral shifts of ocean floor, i.e. movements of hundreds of kilometres. This is quite fascinating because it means that the oceanic crust resembles that of the continents in that it too behaves in a rigid fashion. On land, transcurrent faults, i.e. those in which movement is lateral rather than vertical, are well known, e.g. the San Andreas in California and the Alpine in New Zealand.

All this makes the idea of continental drift not only possible but probable.

1.1.3. Sea-floor Spreading
In the 1960s, a further advance of tremendous importance took place when Harry Hess got the idea, a brilliant one at that, that mid-oceanic ridges are underlain by the hot and ascending limbs of convection cells in the mantle and that the sea floor is being conveyed *away* from the axes and eventually under marginal trenches by the cool and descending limbs (Fig. 1.9). Consequently, the ocean floor must be young—it is undergoing a process of continual renewal.

The oceanic crustal layer is believed to be composed of serpentinised peridotite, effectively the hydrated surface of the mantle. Serpentine is actually about one-quarter water by volume and this could originate in degassing of the rising column of a mantle convection cell. Our layer 3 now becomes serpentinised peridotite, but the sea-floor-spreading hypothesis does not depend on this and layer 3 might equally well be the basalt or gabbro alluded to earlier. An obvious inference of the hypothesis is that it can constitute a mechanism to explain the lateral displacement of

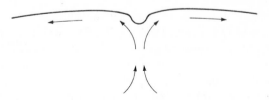

FIG. 1.9.   Sea-floor spreading from the ridge axis.

continents. Lack of a mechanism was always a criticism of Wegener's ideas and this new approach eliminates many difficulties. The continents are seen as riding conveyor belts, not grinding through sima as was formerly thought.

J. Tuzo Wilson advanced matters further by introducing the concept of plates.

1.1.4. Plate Tectonics

In 1965, Wilson proposed that the mobile belts of the Earth, i.e. mountain ranges, island arcs, mid-oceanic ridges and major faults characterised by earthquakes and volcanicity, interconnect and divide the planetary surface into several large and rigid plates (Fig. 1.10). He suggested that any feature at its apparent termination can be transformed into another type.[5]

Another new approach was presented by F. J. Vine and D. H. Matthews.[6] This resulted from detailed mapping of anomaly patterns on the Carlsberg Ridge and its flanks in the northwestern part of the Indian Ocean. Prior to this, the suggestion had been made that the geomagnetic field had undergone several reversals within the last few million years. Now the computed magnetic profiles assuming normal magnetisation of the ocean floor turned out to have little resemblance to the actually observed profiles. In consequence, a model was proposed with half of the crust reversely magnetised in alternating bands. This depends upon the following logic. Assuming that both sea-floor spreading and magnetic reversals take place, basalt magma might well be expected to well up at the axis of the

oceanic ridge, become magnetised in the direction of the magnetic field as it
cooled below the Curie temperature to form a dyke and spread laterally
away from the axis. Repetition of this process with the dyke being split
axially will result in the creation of a series of blocks of alternately normal
and reversed magnetised material aligned parallel to the axis of the ridge
and becoming progressively *older* with increasing distance from this. Using

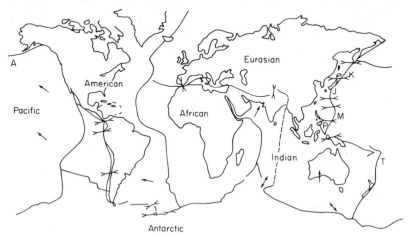

FIG. 1.10.   The major plates of the Earth. Each is believed to move uniformly and
independently so that the geological activity of the planet is mainly restricted to the
areas where these plates are interacting. A = Aleutians; J = Japan; K = Kuriles;
M = Marianas Trench; P = Philippines; T = Tonga Trench.

this model, the computed profiles agreed quite well with observations. This
Vine–Matthews hypothesis constituted a potential magnetic tape recorder
which could be utilised to find out the velocity of the conveyor belt *if* an
accurate time scale for magnetic reversals could be supplied. This came as a
result of the application of potassium–argon dating techniques to lavas and
the construction of a time scale going back about three million years. In
fact, this agreed well with the sequence of reversals established
independently by the study of the magnetism of deep sea sediment cores.[7]

Rates of spreading of the ocean floor were calculated and they vary from
about 1 cm/year on the ridge southwest of Iceland to 4·5 cm/year in the
South Pacific.[8] After this work in 1966, the concept of the lateral mobility
of continents began to be taken very seriously indeed. An intensive
survey of magnetic anomalies over the world's oceans was carried
out by researchers at the Lamont Observatory. The JOIDES (Joint

Oceanographic Institutes Deep Earth Sampling Programme) venture involving the drilling ship *Glomar Challenger* included a traverse across the South Atlantic at roughly 30 °S and drillings were made as far as the Upper Cretaceous and reached basaltic basement often. The age of the sediments directly overlying this basement was found to increase systematically away from the axis of the Mid-Atlantic Ridge by examination of microfossils. A spreading rate of 2 cm/year was inferred. This is important not only because it concurs with work done independently but also because it implies that the rate of spreading may have been more or less constant for a very long time.

1.1.5. Earth History

Wilson first used the term 'plates', but W. J. Morgan extended the transform fault idea to a spherical surface and, in fact, divided the planetary surface into 20 blocks of various sizes divided by three kinds of boundary, namely

1. oceanic rises (zones of crust creation),
2. trenches (zones of crust destruction),
3. transform faults (where crust is neither created nor destroyed).

The blocks were assumed to be perfectly rigid and, as the crust is too thin to show the required strength, they were believed to extend about 100 km down to the low-velocity layer of the mantle. These blocks are plates.

Morgan termed the relatively rigid region of the upper 100 km the tectosphere, but it is more widely known as the lithosphere (Fig. 1.11).[9] He found factual support for a change in spreading rates down the Atlantic as well.

Le Pichon devised a simpler model than Morgan's with only six major plates, namely the American, Eurasian, African, Indian, Pacific and

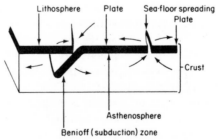

FIG. 1.11.   Plate tectonics: crustal features.

Antarctic.[10] He used two independent sets of data to determine centres of rotation, the spreading rates found from magnetic anomalies and the azimuths of transform faults at the intersection with oceanic ridge axes. He demonstrated how the opening up of the South Pacific, North Pacific, Arctic, Atlantic and Indian oceans can be individually described by a single rotation. Also, the plates do seem to behave to a first approximation as rigid bodies.

The idea that oceanic spreading results from a big increase in the radius of the Earth over the past 200 million years is rejected. Obviously, therefore, any crust formed by spreading from the ridges must be consumed somewhere else. Simultaneously with Le Pichon's work, three Lamont seismologists pointed out that the new global tectonics were much more successful in accounting for earthquake phenomena than earlier ideas.[11]

Some special problems in Earth history may be considered.

(a) Reuniting Pangaean components

The best fit for the South Atlantic has been found to be at the 1000-m contour.[12] It is slightly better than the 2000-m contour.

To fit Africa, North America and Europe, an anticlockwise rotation of Spain (opening up the Bay of Biscay) is postulated. Iceland is thought to be no older than Tertiary and underlain by oceanic crust, and the Rockall Bank is believed to be a submerged part of the continental crust. The good fit in the Atlantic is based on Euler's theorem and confirms the anticipated resistance to distortion of the lithospheric plates. Fitting together South America/Arabia/Africa, Antarctica/Australia and India without leaving gaps is difficult. Perhaps India ought to be fitted to Australia, although other possibilities exist.

(b) Explaining how mountain belts originate

These linear or arcuate features usually comprise deformed sediments of the flysch type sometimes with associated submarine pillow lavas or welded tuffs, and folding and thrusting suggest compressional forces normal to their lengths. Paired island arc and trench systems possess seismicity and volcanicity and resemble mountain belts which indeed they may later develop into. Two types of mountain-building have been distinguished and they are consequent upon plate movements. The first is the island arc–Cordilleran type developing on leading plate edges above subduction zones (Fig. 1.12). Subduction zones are sometimes called Benioff zones and they represent regions of crust destruction as opposed to constructive zones of crust creation and conservative zones where neither event is occurring.

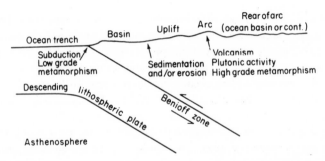

FIG. 1.12.   Cross-section of the crust in a region of subduction (the descending
lithospheric plate undergoing destruction) showing the Benioff zone.

The second is the collision type and forms after impact of continent on
continent or island arc (Fig 1.13). The Circum–Pacific and Alpine–
Himalayan mountain groups are related to the movements of plates
which triggered the disintegration of Pangaea and it is believed that older
mountain belts formed similarly. It has been proposed that the early
Palaeozoic Caledonian mountain belt of northwestern Europe and eastern
North America represents the line of closure of an ancient ocean. Likewise,
the Urals could be a similar late Palaeozoic closure line between Siberia and
Europe. Evidently, Pangaea did not exist earlier than 300 million years ago.
Orogenic belts clearly afford evidence for the formation and disappearance
of oceans through geologic time.

(c) *Accounting for igneous activity*
    This can easily be done in terms of plate tectonics. Volcanic eruptions in
Hawaii have been found to be preceded by minor seismic shocks arising in
the upper mantle, the probable source of the lava. A heat source from below
may be provided by an ascending plume of hot mantle material under either
a ridge or a hot spot accompanied by a local tensional region in the crust.
Basaltic lava eruptions on continents also require an upper mantle source
and overlying tension in the crust.

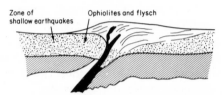

FIG. 1.13.   Continental collision.

The origin of andesitic volcanic rocks and batholiths (granitic intrusions of vast dimensions) are more difficult to explain, however. The former may arise by differentiation of descending slabs of lithosphere, fusion heat being provided by friction with an overlying plate. As such slabs go down, basalt eventually converts to eclogite under increasing lithostatic pressure and this will facilitate the downward pull because of a higher density than peridotite. The content of potash has been found to increase directly with the depth of the seismic zone in present-day island arcs and this most likely relates to different crystal melt equilibria at different pressures and temperatures. Varying contents of potash are also traceable over antique island arcs (represented today by batholiths) and may give a clue regarding the local depth of the Benioff zone.

All clastic sediments are derivable from such igneous rocks. Since granites and andesites may derive from oceanic lithosphere, continent-building is explicable, the presently tectonically stable shields having been island arcs and trenches in previous times such as the Palaeozoic or the Precambrian.

(d) Sea-level changes

Heat flow has been found to decrease systematically with increasing age of the ocean floor away from ridges and, as these latter are isostatically compensated, their relatively high relief may be due to thermal expansion of mantle material. This would fit in with the lower seismic wave transmission velocities occurring beneath the ridges.

It has also been shown that the existing sea-floor topography is in good correspondence with a theoretical model based on heat-flow data and estimated thermal expansion of mantle material. JOIDES borehole data indicate progressive subsidence of the oceanic floor away from ridge axes. Some subsidence in the Indian and Atlantic ocean margins is probably due to cooling of the oceanic floor. On the other hand, world-wide marine transgressions may relate to uplift of oceanic ridges during continental drift.

(e) Ancient ice ages

There is excellent evidence as regards these, e.g. in the Lower Carboniferous when the palaeomagnetists state that the pole was in the Transvaal, glaciers radiated from a number of centres in southern Africa. In the Lower Permian, the pole approached Australia and glacial centres disappeared everywhere but here, Antarctica and a small area in southwest Africa. A major Ordovician glaciation has been detected in the Sahara, confirming the palaeomagnetists' estimate of a pole position located then in North Africa.

Most interesting is Maurice Ewing's and W. L. Donn's hypothesis that ice ages result from movement of the poles into thermally isolated positions. For instance, the equable climates of the Mesozoic confirmed by palaeotemperature measurements partly effected by the writer can be attributed to pole positions in the open ocean. Thus, no permanent ice cover could form even if the pattern of solar radiation was the same as that in the Pleistocene.[13]

From the foregoing detailed discussion, it may be seen that the revolution in geology has had a profound effect and it is relevant now to relate it to investigations into geothermal energy.

1.2. PLATE TECTONICS AND GEOTHERMAL ENERGY

Summarising, it may be stated that the planetary crust, about 5 km thick under oceans and as much as 40 km under continents, is underlain by the mantle passing down into the core. Seismic discontinuities such as the Moho occur and probably are due to compositional changes.

The recent new concept of the outermost few hundred kilometres of the Earth is the plate tectonic model in which the planetary surface, including the sea floor, is divided into several rigid plates in motion relative to each other. These plates comprise lithosphere, including oceanic or continental crust or both, overlying and combined with the uppermost part of the mantle. Oceanic lithosphere is 75–100 km thick and continental lithosphere is about 150 km thick. Beneath the lithosphere lies the so-called asthenosphere which is 400–600 km thick. Although its composition is uncertain, seismic data indicate a region of partial melting in the upper portion and a number of probable density transitions in the lower portion. Along the extensive belt of oceanic ridges, the plates are separating at a rate of a few centimetres annually and new mantle material (magma) is filling the resultant gap. Naturally, in the direction of plate motion away from the ridges, plates must converge and one will sink (subduct) beneath another. Deep oceanic trenches form at these boundaries. Beyond the trenches are volcanic arcs accompanied by shallow to deep seismicity. Boundaries of this type are exemplified by Japan, Indonesia, Kamchatka and the Aleutians as well as the Andes of South America.

When plates converge, both having a layer of continental crust, this is less dense and cannot sink. Such boundaries are characterised by thrust

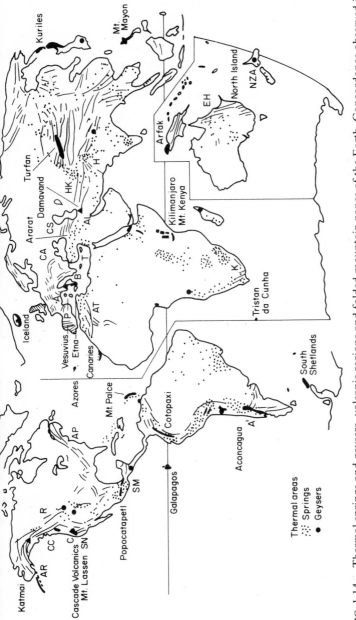

FIG. 1.14. Thermal springs and geysers, volcanic regions and folded mountain zones of the Earth. Concordance observed in many instances relates to plate tectonic considerations. There is also excellent agreement with the major earthquake belts of the planet, especially that circling the Pacific Ocean and often referred to as the Pacific Girdle of Fire. Folded mountain regions: A = Alps; A′ = Andes; AL = Alborz; AP = Appalachians; AR = Alaskan Ranges; AT = Atlas; B = Balkan Ranges; C = Cascades; CA = Carpathians; CC = Canadian Coastal; CS = Caucasus; EH = Eastern Highlands; H = Himalayas; HK = Hindu Kush; K = Karroo; NZA = New Zealand Alps; P = Pyrenees; R = Rockies; SM = Sierra Madre; SN = Sierra Nevada; T = Taurus. Volcanic zones named on map.

faulting, folding and crustal thickening as, for instance, in the mountain regions of the Himalayas and Alps.

At some plate boundaries, neither subduction nor spreading occur, plates of this type sliding past each other along the vast fractures known as transform faults. The most famous is the San Andreas fault system connecting the East Pacific Ridge which enters the Gulf of California to the Gorda Ridge lying off the Oregon–California coast, thus marking the boundary between the Pacific and North American plates.

The spreading of plates is largely confined to oceanic ridges lying in the deep ocean. The cases of the Red Sea and the Gulf of California are atypical because they are only a few million years old and deep oceanic conditions have not yet been reached. A very unusual case is the huge East African Rift Valley because here separation is taking place within continental lithosphere. Actually, the continued spreading of this rift will probably split the African continent and produce new ocean floor. The driving mechanism for plate motion is not fully understood, but seems to be connected with convective movements in the mantle. The internal heat of the Earth is the source of energy. Abnormal terrestrial heat flows occur along the spreading and converging plate boundaries. The mass transfer of heat by magmas generated from the mantle carries heat to the shallower levels of the crust. From such heat sources, geothermal systems develop. All prospective high-enthalpy geothermal regions of the planet, therefore, are to be found within belts of geologically young volcanism and crustal deformation produced by lithospheric plates in motion (Fig. 1.14).

REFERENCES

1. Cheremisinoff, Paul N. and Morresi, Angelo C. (1976). *Geothermal Energy Technology Assessment*. Technomic Publishing Co. Inc., Westport, Conn., 164 pp.
2. 'Around the world alternate energy is sought' (1975). *New York Times*, January 26 issue, p. 23.
3. White, D. E. and Williams, D. L. (1975). Assessment of geothermal resources of the United States—1975. *US Geol. Surv. Circ.*, **726**.
4. Bowen, Robert (1971). The new global tectonics. *Sci. Progr. (Oxford)*, **59**, 369–88.
5. Wilson, J. Tuzo (1965). A new class of faults and their bearing on continental drift. *Nature (London)*, **207**, 343–7.
6. Vine, F. J. and Matthews, D. H. (1963). Magnetic anomalies over oceanic ridges. *Nature (London)*, **199**, 947–9.

7. Cox, A., Dalrymple, G. B. and Doell, R. R. (1967). Reversals of the Earth's magnetic field. *Sci. Amer.*, **216**, 44–54.
8. Vine, F. J. (1966). Spreading of the ocean floor—new evidence. *Science*, **154**, 1405–15.
9. Morgan, W. J. (1968). Rises, trenches, great faults and crustal blocks. *J. Geophys. Res.*, **73**, 1959–82.
10. Le Pichon, X. (1968). Sea floor spreading and continental drift. *J. Geophys. Res.*, **73**, 3661–97.
11. Isacks, B., Oliver, J. and Sykes, L. R. (1968). Seismology and the new global tectonics. *J. Geophys. Res.*, **73**, 5855–99.
12. Bullard, E. C., Everett, J. E. and Smith, A. G. (1965). The fit of the continents around the Atlantic. *Phil. Trans. Roy. Soc. London, Ser. A*, **258**, 41–51.
13. Bowen, Robert (1966). *Palaeotemperature Analysis*. No. 2 in the series *Methods in Geochemistry and Geophysics*. Elsevier Publishing Company, Amsterdam, London, New York, 265 pp.

CHAPTER 2

Geothermal Systems

The origin of geothermal energy has been shown to be the mantle of the Earth. It may be useful now to look at the history of geothermal systems.

2.1. HISTORY

Unexpectedly, the value of geothermal energy has been recognised by Man for thousands of years if only in the form of hot springs, waters of which have been heated by hot rock and thereafter ascend to the surface of the planet. The Romans and the Greeks certainly used these both for recreation and therapeutic purposes and their resorts were scattered throughout the Mediterranean area and as far away from that as the British Isles. J. Barnea and J. J. O'Brien indicated that the ancient Japanese and other peoples in the Far East also employed hot springs as medicinal centres.[1,2] In more recent times, that area known as the 'gates of Hell' was first noted by W. B. Elliott in 1847 and is now termed The Geysers. It is to be found in Sonoma County, California, about 90 miles north of San Francisco and may well comprise the largest accumulation of geothermal steam in the world. When the Gold Rush began, the entire region became a resort with hot mineral springs and the development of the energy source had to wait until the twentieth century when it certainly occurred.[3]

Of course, such health spas exist in many other countries such as France, Italy, Austria, Russia, China, Indonesia, the Philippines, New Zealand and elsewhere. One of the most famous geothermal regions is at Larderello in Tuscany, Italy, and boric acid and other chemicals were extracted from the steam jets there as early as the 1700s, probably the first, as it were, industrial application. In fact, the first geothermal utilising body was the Larderello

Company which was operating in 1928. The company was owned by a Florentine prince, Ginori-Conti, and appears to have been the first to have the idea of employing natural steam for the generation of power. Actually, the Larderello field began producing electricity in 1904 and subsequently ever more steam wells were made, thus enlarging the electrical output until today the field has a 400-megawatt (MW) capacity. It is hardly surprising, therefore, that Italians are foremost in the world today in geothermal energy matters. In fact, they are active outside Italy also, for instance in Iran as the writer observed during his stay in Tehran in connection with a geothermal energy development project during 1978. It is strange but true that the USA was not especially rapid in developing its undoubtedly vast resources of geothermal energy whereas the Italians are continually forging ahead. Innovation is a significant feature. It has been noted earlier that earlier attempts to exploit Larderello entailed both electrical power production and boron extraction. Heat exchangers were used so that a clean fluid could be employed in the turbines but later on, turbines more resistant to corrosion and abrasion came in and plants utilising intermediate heat exchangers were replaced by direct intake turbines which were cheaper and obviated losses at the heat exchangers. As a result, more power resulted per unit of steam.

Additional innovations in Italy include the following:

1. Installation of rather small (1·5–5 MW) back pressure turbines exhausting directly to the atmosphere and employed on individual wells in the initial stages of development of a new field. These are extremely useful because they are capable of handling steam with a high content of non-condensable gases which may rise to as much as one-third by weight and include carbon dioxide. The presence of these gases, accumulated over a long time interval in the upper parts of a geothermal reservoir, is deleterious to exploitation. By the technique described, they can be released and thus the ratio of non-condensable gases to steam may be altered until usage in conventional condensing turbines becomes feasible.

2. The above technique has an additional attraction in that it enables the temperature–pressure–volume relationships of the reservoir to be determined by production testing and it is possible to make predictions regarding the life span of a reservoir before any money is invested in larger scale development. Of course, electricity produced in the meantime can be sold and revenues so derived devoted to exploitation.

3. At Larderello, the technique of employing air as the drilling fluid in the drilling of geothermal wells has facilitated penetration into low-pressured steam zones without their becoming sealed off with a mud cake as would be the result of utilising conventional mineral-bearing muds as circulation fluids. In fact, it is now a normal practice to use air instead of mud in all vapour-dominated geothermal systems and the approach has been found to increase greatly the flow per well (because more zones are able to produce steam).

Turning now to the USA, the first efforts there to utilise geothermal power were made in the 1920s when eight steam wells were drilled at The Geysers and the steam utilised in order to power a generator driven by a small reciprocating engine. Availability of hydroelectric power cheaply plus a lack of demand in the immediate vicinity of the geothermal field made the field an economically unattractive proposition at that time. However, later, specifically in 1955, the development of The Geysers was again attempted and this time led to the building of the first power plant in 1960. This has a capacity of 12·5 MW. Later drilling and construction resulted in the existing steam production capacity of approximately 1000 MW and operating plants of 502 MW.

Other areas in the USA have proved fruitful in the production of geothermal energy. One of these is located a few miles to the north of the Salton Sea and possessed a steam yield considered inadequate for the generation of power. Many mud pots exist and there is considerable fumarolic activity. Drillings were made for carbon dioxide as early as 1932, and between 1934 and 1954 more than $2·5 \times 10^9 \, ft^3$ of this gas were derived from 65 wells. Actually, on the field there operated a dry-ice plant and this confirmed the commercial importance of the geothermal resource.

Another area in the States which was surveyed in 1929 and 1930 is the Yellowstone National Park where study wells were sunk in the Upper Geyser Basin and the Norris Basin, thereafter being sealed when data had been amassed. Additional locations include Mount Katmai in Alaska and Coso Hot Springs in California.[4]

Much of the early exploratory work was privately undertaken and this hampered development to an extent because, quite naturally, investors require returns on capital outlay! After the second world war, of course, the pace of geothermal research and development quickened and significant advances began to take place.

2.2. GEOTHERMAL SYSTEMS

In Chapter 1, the ever increasing urgency of the search for new and preferably non-pollutional sources of energy was indicated and this is particularly applicable to the USA. Here is demonstrated the planetary imbalance in energy consumption. According to the world energy statistics of the UN, Americans consume one-third of the world's energy production although constituting only about 6% of the population of the Earth! Leaving aside the developing countries, the USA accounts for about twice as much energy consumption *per capita* as the Federal Republic of Germany or the United Kingdom and four times as much as Japan! That it can do so, of course, has been due to the availability until now of low-cost energy in the form of abundant domestic fossil-fuel reserves. However, these are running out fast, a fact startlingly reflected in the transformation of the USA in the late 1960s into a net importer of petroleum products (now importing over a third of its needs). This, together with a shortage of storage capacity and delay in provision of nuclear reactor electric-generating capacity, has resulted in an inability to provide new capacity commensurate with the growth in energy consumption. An unpleasant consequence has been power shortages, especially in the extremes of summer and winter weather conditions. Clearly, the USA has changed its status from self-sufficiency in energy to dependence on other countries to meet its requirements. Starting from a lower base, the needs of the west outside America will probably increase at a rate of 7–8% annually compared with the 4·2% rate in the States. Since no industrialised country is now self-sufficient, a lively competition for available supplies may be forecast between the USA and the free world. The significance of the utilisation of geothermal energy in the USA is apparent and this may account for W. J. Hickel's estimate that, by 1985, America might be producing at least as much as 132 000 MW of electric power from geothermal reserves.[5]

As seen above, the Earth's inner heat is a vast energy resource, in fact one that greatly exceeds that which could be provided if all the uranium and thorium in the crust were used in breeder reactors. Sadly, most of it is too deep or too diffused to be detected superficially where it may be swamped by incident solar radiation. Manifestations of it do occur, however, in the regions determined by plate tectonic considerations in the form of volcanoes, geysers, thermal springs, fumaroles, etc., all indicating that our planet is a huge heat engine. Eruptions from volcanoes can emit materials characterised by temperatures as high as 800 °C.

Employment of the heat of the Earth depends upon a number of factors. One is enthalpy, heat content, but this is not always as relevant as it might appear to be. Molten lava obviously possesses an extremely high enthalpy, but it is, as far as present technology is concerned, practically useless as a source of geothermal energy. This is not to say, of course, that a suitable methodology will not be devised. Thus, D. L. Peck indicated that 42 inferred molten bodies of silicic and intermediate composition with total heat energy at least 30 times the estimated heat content of all hydrothermal systems in the USA at depths of less than 3 km have been identified.[6] Undoubtedly, therefore, at least in America and certainly elsewhere as well, molten igneous systems represent a sizeable portion of the geothermal resource base. Consequently, better drilling technology must be devised to enable drilling to enter the magma at depths of 3–6 km (pressures of 1–2 kbars) and temperatures of 650–1200 °C. At present, drilling can only be effected at 1–2-kbar pressures relevant to this discussion at temperatures below 250 °C. Also, a durable and efficient heat-extraction system is fundamental, namely one which would function for at least a generation. Sandia Laboratories have investigated several possibilities. One is a closed system with a long heat-exchanger tube in the magma, water for steam generation as a working fluid and a conventional turbine generator. Another involves the use of gas as such a fluid and solid-electrolyte fuel cells to increase the efficiency of energy extraction. Obviously, field tests could be carried out, for instance in a Hawaiian lava lake. The feasibility of magma-energy utilisation is connected with the heat-transfer coefficient of magmas of different compositions. Thus, H. C. Hardee calculated that, using a long heat-exchanger tube with water as the working fluid, long-term heat-extraction rates from basaltic magma were only 1 kW/m^2 for non-convecting magma, but orders of magnitude higher for vigorously convecting magma.[7]

As regards hot, dry rock, this has been found to be of use only in a very localised and limited manner.

As seen in the preceding chapter, high-enthalpy geothermal systems occur in regions of recent orogeny, volcanicity and rifting, thus relating to plate tectonic considerations. All over the world, temperatures increase with depth in boreholes, wells and mine shafts, usually at a rate of just over 0·5 °C per 100 ft of depth. In the above-mentioned areas, of course, anomalously high gradients occur. Curiously, such geothermal prospects have so far only been exploited in highly developed countries.

As well as temperature gradient, the thermal conductivity of rocks is very important in determining heat flow. The thermal conductivity is defined as the quantity of heat transmitted through a 1-cm^2 surface in 1 s under a thermal gradient of 1 °C/cm. Some typical thermal conductivities are listed in Table 2.1.[8]

Where heat flow is effected by thermal conduction, amounts leaving the surface of the Earth per unit area are products of the thermal conductivity and the temperature gradient. Field measurements of these parameters are essential. Naturally, they depend upon ground-water flow, porosity,

TABLE 2.1

TYPICAL THERMAL CONDUCTIVITIES OF VARIOUS ROCK TYPES

Rock type	Thermal conductivity $(10^{-3} \, cal/cm \, °C \, s)$
Granite	6–9
Dolerite	7–8
Gneiss, normal to foliation	5–9
Gneiss, parallel to foliation	6–11
Quartzite	7–19
Limestone	4–7
Dolomite	9–14
Sandstone	4–11
Shale	3–6
Rock salt	13–17
Wet ocean sediments	1·7–2·4

mineralogy and the water content of pores. A consequence is that near-surface temperature gradients cannot safely be projected downwards since rock conductivity may well change in that direction.

It was noted above that temperature gradients are anomalously high in active orogenic, volcanic and rifting regions of the planet and it is in no way surprising, therefore, that, for instance in the Hungarian plain, wells have been drilled and showed temperature increases as great as 1 °C in only 7 m.

Heat flows vary with regard to both time and space, and some mean values from different geological provinces are shown in Table 2.2.[9] Of course, these figures must not be taken as especially accurate. They cannot, therefore, be employed as material suitable in identifying geothermal areas.

Technical factors militate against accuracy. For instance, during drilling, bits generate heat. Also, subterranean hot waters contribute heat and may

TABLE 2.2

HEAT FLOW FROM DIFFERENT GEOLOGICAL PROVINCES

Province	Heat flow $(\mu cal/cm^2\,s)$
Precambrian shields	$0\cdot92 \pm 0\cdot17$
Palaeozoic mountain regions	$1\cdot23 \pm 0\cdot40$
Continental plates	$1\cdot54 \pm 0\cdot38$
Post-Palaeozoic mountain regions	$1\cdot92 \pm 0\cdot49$
Mid-oceanic ridges	$1\cdot82 \pm 1\cdot56$
Deep ocean basins	$1\cdot28 \pm 0\cdot53$
Mean for the entire Earth	$1\cdot50 \pm 0\cdot15$

remove heat from one place to another. It is possible that oceanic measurements are more reliable than those effected on continents.

Nevertheless, G. Sestini believes that if accurate heat-flow values are available in reconnaissance studies of geothermal areas, they can facilitate reduction in the number of wells which later have to be drilled during the exploitation stage.[10]

Examining geothermal fields will involve

1. relating geological structures to heat flow and temperature gradient;
2. determining the influence of well sites;
3. assessing the heat-flow values at different depths in wells drilled;
4. finding out the effects of the diameters of wells and their filling up systems on measurements made;
5. approximation of determined heat-flow values by means of temperature gradients and thermal-conductivity measurements.

It will be useful now to attempt to classify geothermal systems. Five major types of geothermal energy resource with distinctive and differing characteristics have been recognised. These are as follows:

1. Hydrothermal convective systems containing relatively high-temperature water at shallow depths. They may be either
   (a) vapour-dominated at a characteristic temperature exceeding 150 °C and capable of producing superheated steam, or
   (b) liquid-dominated, ranging from below 90 °C to more than 350 °C and capable of producing a mixture of liquid and vapour or hot liquid only.

2. Geopressurised resources. These are pressurised water reservoirs within sedimentary basins capable of supplying both heat and mechanical energy together with dissolved methane.
3. Hot, dry rocks. Non-molten, but very hot rock structures with insufficient water to be considered a hydrothermal convective system. The temperature is usually less than 650 °C.
4. Normal gradients. Conduction-dominated areas produced by heat flows, radiogenic heat production and thermal conductivity of rocks showing temperature ranges from 15 °C to approximately 300 °C within the first 10 km of the crust.
5. Magma. Molten rock at temperatures exceeding 650 °C.

As mentioned earlier, some of these have little practical significance at the present time, but may become exploitable given improved technology in the future.

In geothermal systems capable of exploitation with existing technology, ground water is a basic component. Evidence from recent isotope analyses demonstrates that most water in known geothermal systems is of meteoric origin, i.e. the water is part of the normal ground water participating in the hydrologic cycle of the Earth. A small amount, however, perhaps less than 10%, does derive directly from magma (juvenile water).

If this water is considered, it becomes apparent that as it is heated, its density will decrease. If the overlying rock is permeable, then a convection cell arises. Clearly, an impermeable (cap) rock must be present also in order to prevent escape of the fluid to the surface. The thermal gradient in this will be high, decreasing rapidly in the convectional geothermal system's upper parts (where convection will be most prominent). Thereafter, the temperature will show little variation with depth and is termed the base temperature. It is in this part of the system that the reservoir occurs. Of course, leaks from the reservoir to the surface manifest themselves in such phenomena as steam vents, geysers and fumaroles.

Simplifying the above classification of geothermal systems on a pragmatic basis, two general classes of exploitable character may be derived and these are

1. vapour-dominated systems, and
2. water-dominated systems.

Vapour-dominated systems produce, as noted above, saturated to slightly superheated steam, temperatures often being around 250 °C at pressures of

30–35 bars. Generally, the associated reservoirs comprise fractured or porous rocks and the flows from wells may range from a couple of thousand kilograms per hour to as much as a quarter of a million kilograms per hour from depths of up to 2500 m. Non-condensable gas contents of the steam can be as much as a third, but rapidly decline with production until a stable figure as high as 5 % may be attained. The much greater initial contents imply previous accumulation in the reservoir. The reservoirs usually are below hydrostatic pressure indicating that they are sealed off from the infiltration of ground water and probably developed from high-temperature, liquid-dominated systems which sealed off their cooler margins through geologic time by precipitating dissolved material, primarily silica.[11] Additional slow escape of water creates space for steam and a deep liquid phase, perhaps a hot brine. The heat, of course, comes from below, most likely from a magmatic intrusion. Many of the world's most famous geothermal fields belong to this category, for instance the steam fields at Larderello and The Geysers and also that located at Matsukawa, Japan, and the reservoir characteristics for all of them are quite similar. At Larderello, porous limestone and dolomite constitute the reservoir rocks compared with indurated, fractured graywacke sandstone and volcanics at The Geysers and fractured volcanics in the Japanese case. Lithological variations, of course, are not significant in comparison with resemblances in porosity and/or fracturation state, factors which permit entry and circulation of waters.

Liquid-dominated systems may be divided into two subtypes, namely

1.   that possessing high-enthalpy fluids exceeding 200 cal/g, and
2.   low-enthalpy, fluid-bearing subtype below this value.

The practical utility of this scheme is that recognition of them enables fluids useful for generation of electric power to be separated from those which are useful for other purposes.

There is a significant physical distinction to be drawn between vapour-dominated and liquid-dominated hydrothermal convective systems. In liquid systems, the reservoir pressures are near hydrostatic pressures (around 0·1 bar per metre depth). This means that, at depths of between 1500 and 2500 m, pressures are 100–250 bars in contrast to 30–35 bars found in the vapour-dominated systems.

1. High-enthalpy fluid systems contain waters having dissolved solids ranging from 2000 ppm to over 250 000 ppm, with temperatures going as high as 388 °C.[11]

The main anion of the dissolved solids is chloride with smaller quantities of sulphate and carbonate. The main cations are sodium and potassium with smaller quantities of calcium and magnesium. As much as 800 ppm of silica may occur as well as lesser amounts of fluoride, boron, etc., all of which are a nuisance. Wells drilled into such systems yield a water–steam mixture and, of course, the steam may be separated at an appropriate pressure and utilised in order to operate a turbine. The amount of non-condensable gas does not normally comprise more than 1 %. A famous high-enthalpy, liquid-dominated system is located at Wairakei, New Zealand. Here, wells are drilled into a permeable volcanic rock which is capped by impermeable sediments and the temperature of the fluid is about 260 °C. Some one-fifth is flashed to steam for the production of power.

2. It is known that low-enthalpy, liquid-dominated systems show a greater variability in properties and in some cases the sulphate ion may predominate, others having predominant carbonate–bicarbonate. Salinities show a tendency to be lower, some being actually potable. The dissolved silica content, being a function of temperature, is less and such toxics as fluorine and boron occur in smaller amounts. The temperatures range from as low as 10 °C above average annual temperature to an arbitrary 200 °C.[11] They fall into the wet-fields category of Christopher and Armstead's hyperthermal fields.[12]

Under low-conductivity cap rocks sometimes occur low-enthalpy waters in deep sedimentary basins with temperatures between 50 and 120 °C. Such reservoirs may be considerable in extent and the Hungarian Basin is an example. Here and elsewhere, there does not appear to be any evidence of recent volcanism and it is believed, therefore, that the source of heat in such cases may well be a terrestrial heat flow which is somewhat more than normal. Probably the best-known region of low-enthalpy reservoirs is Iceland and these are extensively utilised for space heating. In fact, the government is proposing to expend $US4·3 million of their budget on geothermal exploitation during the decade 1973–83.

2.3. PLANETARY HEAT

The association of geothermal resources with the concept of plate tectonics has been indicated and the fact that mantle-derived magma is a primary source of terrestrial surficial and crustal heat noted. The molten origin of the Earth has been out of favour ever since Harold C. Urey and others demonstrated that our planet probably originated by accretion of cold matter of dust particle to asteroid size range over a relatively short time interval. This perhaps 4·6 aeons (i.e. 4·6 × 10⁹ years) ago. At this stage, the interior was most likely a cool solid. Later, as accumulation continued, heating due to impact began and surficial melting may have occurred. H. W. Menard suggested that, if the Earth accumulated in a couple of million years, then the maximal heating took place when the radius was between a quarter and a half of its present value.[13] This, in fact, corresponds to the liquid region of the existing core. The mantle overlying this would be both solid and cool, but it must be remembered that

1. most of the mantle will have been melted several times by superimposed impacts,
2. affected parts will have solidified and cooled between such impacts,
3. sooner or later, some of them must have been so deeply buried that they were no longer susceptible to additional melting.

As the mantle accumulated, low-density materials were carried upwards, most probably to comprise the continents.

Thus, however the Earth originated, it seems to have acquired heat at an early stage in its development and this is the basis of the natural heat alluded to at the beginning of this book, i.e. that heat which passes from the interior to the surface at around 1·5 μcal/cm² s (some people give different figures, cf. 1·2 μcal/cm² s given by Paul N. Cheremisinoff and Angelo C. Morresi[14]). However, the current high heat flux requires an extra contribution and this may well be due to radioactivity.

TABLE 2.3

DISTRIBUTION OF RADIOELEMENTS IN SOME SEDIMENTARY ROCKS

Rock	Uranium (10^{-6} g/g of rock)	Thorium (10^{-6} g/g of rock)
Sandstone	up to 4	—
Quartzite	1·6	—
Clay	4·3	13·0
Claystone and shale	3·0	—
Limestone	1·5	0·5
Dolomite	0·3	—

The relevant radioelements are thought to be uranium, thorium and potassium. To illustrate their distributions in certain sedimentary rocks, reference may be made to Table 2.3.[15]

Variations are due to differences in sorption capacity of the different sedimentary materials for radioelements migrating through them in ground water.

The average value and limits of variation of radioactivity in certain sedimentary rocks may be appended also (Table 2.4).[15]

TABLE 2.4
AVERAGE VALUE AND LIMITS OF VARIATION OF
RADIOACTIVITY IN SEDIMENTARY ROCKS

Rock	$RaEq^a$ $(10^{-12}\, g/g\ of\ rock)$
Anhydrite	0·5
Brown coal	1·0
Rock salt	2·0
Dolomite	0·5–10
Limestone	0·5–12
Sandstone	1·0–15
Clayey sandstone	2·0–20
Clayey limestone	2·0–20
Carbonaceous claystone and shale	3·0–25
Claystone and shale	4·0–30
Potassium salt	10·0–45
Deep sea clay	10·0–60

^{a} RaEq is equivalent radium, i.e. an amount of radium capable of emitting an equivalent radiation dose.

From the above, it is clear that homogeneous clay-free organogenic and quartzose sediments are the least radioactive. Claystones and shales, on the other hand, show an increase in radioactivity which appears almost directly related to increasing clay content.

The three radioelements in question, i.e. ^{40}K and the uranium and thorium series, emit gamma radiation with energies ranging from 0·24 MeV to 2·62 MeV. Uranium has two radioisotopes, namely ^{235}U and ^{238}U, and thorium has one, ^{232}Th. The half-lives and heat productions characterising them are listed in Table 2.5.[14]

The radioelements show increasing concentration in acid igneous rocks. Thus, the uranium content of basalts averages 0·000 109 % and in granites it

TABLE 2.5

HALF-LIVES AND HEAT PRODUCTION OF K, U AND Th
RADIONUCLIDES

Radionuclide	$(10^9 \ years)$	Isotope %	$(cal/g \ year)$
Potassium-40	1·31	0·012	27×10^{-6}
Uranium-235	0·71	0·72	0·03
Uranium-238	4·50	99·27	0·70
Thorium-232	13·90	100·00	0·20

is on average 0·000 442 %. For potassium, the comparable figures are 1·07 % in basalts and 3·7 % in granites. Heat-production figures for these two rock types are given in Table 2.6.[14]

It may be inferred on theoretical grounds that radioactive elements are concentrated in the crust and upper mantle of the planet. If this was not the case, the Earth would be molten. As a consequence of the observed thermal gradient, it may also be inferred that a large part of the mantle is at the melting temperature and liquefied pockets represent locally relieved pressure zones. At the Moho, the temperatures are believed to be anything from 500 °C to 700 °C. For comparison, it is estimated that the central core temperature is probably in excess of 4000 °C.

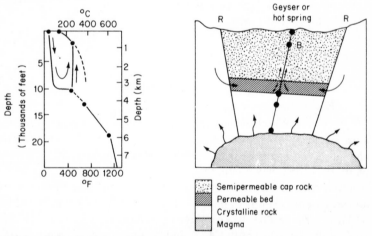

FIG. 2.1. Left: depth versus temperature curve for a typical hot-water system; right: model of a high-temperature, hot-water geothermal system (B being point of commencement of boiling, R points of recharge to ground-water system and arrows indicating emission of heat from magma).

TABLE 2.6
HEAT PRODUCTION IN GRANITE AND BASALT

Rock type	Heat production (cal/g year)			
	Uranium	Thorium	Potassium	Total
Granite	3·4	4·0	0·9	8·3
Basalt	0·44	0·54	0·23	1·21

Both heat sources and geological factors are responsible for the origin of geothermal hot spots. The latter relate to plate tectonics and Fig. 2.1 illustrates some aspects of the matter.

One problem of great interest to geologists cannot be explained, however, and this is whether the Earth is cooling down or heating up. There is nothing which may usefully be said at this stage on the question.

2.4. IMPORTANT KNOWN PLANETARY GEOTHERMAL RESERVES

A summation of important known planetary geothermal reserves may be given and this follows below (countries listed alphabetically).

Algeria: Hamman Mescoutine. Geologically, this resembles the Larderello region in Italy. French oil company exploration is active.

Chile: El Tatio. UN exploration proceeding.

Colombia: There is exploration in the Cordillera.

Czechoslovakia: Famous hot springs occur at Marienbad and Karlsbad as well as in the Carpathians.

El Salvador: A vapour-dominated type commercial field is located at Ahuachapan and it has sufficient steam for 60-MW power production (a first phase of 30 MW exists).

France: Here, there are no less than 1200 hot and warm springs. A space-heating project exists at Melun and a hot-water resource is available in the Paris area.

Hungary: As early as the 1930s, a Budapest district with 2000 inhabitants was heated with hot water from the Danubian Basin. As noted earlier, another such resource occurs in the Pusta.

Iceland: Geothermal energy is widely utilised here.

Iran: One very promising area is that of Mount Damavand, a post-Quaternary and still active volcano with many hot springs in its vicinity. A 50-MW power generation is anticipated. Other potential areas include the volcanoes Sahand and Sabalan as well as the Maku–Khoy region on the Turkish border near Tabriz. The Damavand region is of great significance because it is quite close to the capital city, Tehran.

Italy: The Larderello type, i.e. a vapour-dominated field, is very well known indeed, but development is progressing rapidly in other areas such as Vitergo–Monte Simino. Of course, the famous steam field is still being investigated as also is the Travale field east–southeast of it (by R. Cataldi and others).

Japan: At Matsukawa, a dry-steam area with a capacity estimated at 200 MW exists and the Geothermal Energy Association believes that as much as 10 000-MW capacity may be attained in the country as a whole.

FIG. 2.2.   The geothermal regions of Japan.

Geothermal energy is particularly important in Japan which, it is estimated, will be importing about 90% of its energy needs by 1980. Hence, every effort is being made to develop the geothermal fields and in 1967, a power plant was constructed in the Otake area with a 16-MW capacity. Actually, 10 areas have been selected for investigation, an easy task in a land of active volcanism full of hot springs and fumaroles (Fig. 2.2). As well as industrial employment, geothermal energy is currently being used in spas and agriculture and Japan is supplying power plants for The Geysers in California and Cerro Prieto in Mexico.

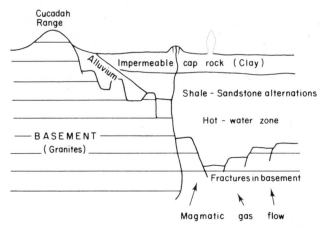

FIG. 2.3. Geological cross-section of Cerro Prieto geothermal field. Adapted from the brochure on this field issued by the Comision Federal de Electricidad, Mexico in 1971.

Mexico: For almost a quarter of a century, exploration and development have been proceeding and four active fields are due for exploitation, the best known being the afore-mentioned Cerro Prieto near Mexicali. Here, intense investigations have been effected and it appears that this is a part of a field in some way controlled by the San Andreas fault network (*v.* Fig. 2.3). The reservoir temperatures exceed 300 °C and the mean well depth is over 4000 ft. A 37·5-MW source is envisaged.

New Zealand: Investigations have been going on for at least 25–30 years, in fact ever since the development of the wet-steam field at Wairakei in the 1950s. Broadlands lies 30 km northeast of this area, also within the Taupo volcanic zone on the North Island, a major volcano–tectonic depression extending 260 km from the active volcanoes of Tongariro National Park to White Island Volcano in the Bay of Plenty. The zone contains almost 11 000 km³ of lava, ash-flow tuff and air-fall tuff (almost all rhyolitic in composition) erupted from the Pliocene until the Holocene. Actually, all of the country's active volcanoes, hot springs and thermal areas lie within this region. Broadlands is one of at least 12 major geothermal areas and was chosen for exploration on the basis of a regional de-resistivity survey. Probably hot water escapes through recently active faults. Proximity to these was a drilling factor and, to 1971, 18 wells were sunk. All but one had temperatures exceeding 270 °C. The potential is 120 MW.

Turkey: At least one major field exists and there are probably at least 10 other geothermal areas, hardly surprising in the country of Mount Ararat.

USA: An enormous number of hydrothermal convective systems are known and may be tabulated summarily as shown in Table 2.7.

TABLE 2.7
CHARACTERISTICS OF HYDROTHERMAL CONVECTIVE SYSTEMS IN THE USA

	Number	Subsurface area (km^2)	Volume (km^3)	Heat content[a] ($10^{18}\,cal$)
Vapour-dominated systems (240 °C)	3[b]	122	194	26
Known hot-water systems				
1. high temperature (150 °C)	63	1 414	2 995	371
2. intermediate temperature (90–150 °C)	224	2 938	4 564	345
Total hydrothermal convection systems	290	4 474	7 753	742

[a] It may be noted that $10^{18}\,cal$ = heat combustion of 690 million barrels of petroleum or 154 million short tons of coal.
[b] These are The Geysers, Mount Lassen (California) and the Mud Volcano System in Yellowstone National Park (Wyoming).

USSR: Geothermal exploration resulted in operation of 11 projects as of 1972 and the government there is planning a tenfold increase during the present decade. It has actually been suggested that over half of the territory of the land is occupied by thermal waters of economic potential.[16] If this is indeed the case, then the geothermal resources reserve could provide an energy total exceeding the total coal, oil, gas and peat reserves of the Soviet Union!

Many other countries possess lesser geothermal possibilities which are of considerable scientific interest and these include the UK (Bath, for instance), Austria (Baden), Greece, Ethiopia, Kenya, China and the Philippines.

It may be appropriate here to add data on world geothermal power-generating capacity for 1972 (Table 2.8).[17]

In regard to these and other areas, the important point to be borne in mind is that all procedures, exploratory and exploitative, refer to one

TABLE 2.8
DATA ON GLOBAL GEOTHERMAL POWER-GENERATING CAPACITY FOR 1972

Country	Field	Electric capacity (MW)			
		Operating	Under construction	Vapour-dominated systems	Hot-water systems
Iceland	Namafjall	2·5			2·5
Italy	Larderello	358·6		358·6	
	Monte Amiata	25·5		25·5	
Japan	Matsukawa	20		20	
	Otake	16		16	
Mexico	Cerro Prieto		37·5		37·5
	Pathe	3·5			3·5
New Zealand	Wairakei	160			160
	Kawerau	10			10
USA	The Geysers	302	110	412[a]	
USSR	Pauzhetsk	5			5
	Paratunka	0·7			0·7

[a] 908 by 1976, v. Chapter 6.

commodity alone, namely *heat*. Heat concentrations are, as seen earlier, absolutely necessary because the diffused heat found everywhere is commercially useless. For instance, for electricity to be generated economically by steam-fed turbines, temperatures in the geothermal reservoir must be in excess of 180 °C. Naturally, other factors such as permeability and the supply of adequate water must also be right.

Some geothermal resources may well not be exploitable even if large, should they lie too deep. So far, the deepest existing geothermal well is at The Geysers and penetrates down to 3 km. Deeper ones could be drilled, but there are limits with presently available technology.

In some cases, where unfavourable factors exist, they may be ameliorated. For instance, if porosity or permeability are unsatisfactory, they may well be improved by artificial means such as blasting or thermal fracturing. The use of nuclear technology is also feasible.

Exploration methodology will be discussed later, but it may be mentioned here that geothermal areas are distinguishable by the occurrence of igneous masses, perhaps 5–10 km deep, the heat from which energises the overlying meteoric circulation (convective) system. An example is The Geysers which involve a Quaternary geothermal system emplaced in Franciscan metasediments and volcanics of Mesozoic age.

REFERENCES

1. Barnea, J. (1972). Geothermal power. *Sci. Amer.*, **226**, 70–7.
2. O'Brien, J. J. (1972). Geothermal resources as a source of water supply. *J. Amer. Water Works Assoc.*, **64**, 694–700.
3. Schuster, R. (1972). Turning turbines with geothermal steam. *Power Engineering*, **76**, 36–41.
4. Wehlage, E. F. (1974). Geothermal energy's potential for heating and cooling in food processing. *Geothermal Energy*, **2**(12), 7–14.
5. Hickel, W. J. (1972). *Geothermal Energy: A Special Report.* University of Alaska, Fairbanks.
6. Peck, D. L. (1975). Recoverability of geothermal energy directly from molten igneous systems. *Geothermal Energy Development.* Hearing before the subcommittee on energy research and water resources of the Committee on Internal and Insular Affairs, US Senate. US Government Printing Office, Washington, D.C., pp. 256–8.
7. Hardee, H. C. (1974). *Natural Convection in a Spherical Cavity with Uniform Internal Heat Generation.* Sandia Laboratories, SLA-74-0089, 20 pp.
8. Bullard, E. (1973). Basic theories. In: *Geothermal Energy*, ed. H. C. Armstead. UNESCO, Paris, pp. 19–29.
9. Gilluly, J., Waters, A. C. and Woodford, A. O. (1968). *Principles of Geology.* W. H. Freeman and Company, San Francisco, 687 pp.
10. Sestini, G. (1970). Heat flow measurements in non-homogeneous terrains. Its application to geothermal areas. *Geothermics (Pisa)*, **2**(1), 424–36.
11. Bolton, R. S., Bowen, R. G., Groh, E. A. and Lindal, Baldur (1977). Geothermal energy technology. Section 7 in: *Energy Technology Handbook*, ed. Douglas M. Considine. McGraw-Hill Book Company, New York, pp. 1–57.
12. Christopher, H. and Armstead, H. (1978). *Geothermal Energy.* E. & F. N. Spon Ltd, London, 357 pp.
13. Menard, H. W. (1974). *Geology, Resources and Society. An Introduction to Earth Science.* W. H. Freeman and Company, San Francisco, 621 pp.
14. Cheremisinoff, Paul N. and Morresi, Angelo C. (1976). *Geothermal Energy Technology Assessment.* Technomic Publishing Co. Inc., Westport, Conn., 164 pp.
15. Feronsky, V. I. (1968). Stratification of aquifers. In: *Guidebook on Nuclear Techniques in Hydrology (Section IV.B.4)*. Technical Reports Series Number 91, p. 157. International Atomic Energy Agency, Vienna.
16. Special Report (1973). *Ground Water and the Geothermal Resource.* Geraghty and Miller Inc., Water Res. Bdg, Manhasset Isle, Port Washington, New York, 14 pp.
17. White, Donald E. (1973). Characteristics of geothermal resources. In: *Geothermal Energy*, ed. Paul Kruger and Carel Otte. Stanford University Press, Stanford, Ca., pp. 69–94.

CHAPTER 3

The Search for Geothermal Resources

As was observed earlier, heat is the phenomenon for which exploration teams are looking. It differs from any other of our sources of energy in that it can be directly utilised and does not require processes such as combustion or fission prior to employment. This is important from the economics standpoint when comparison is made with coal, oil or uranium, for instance. From this, it may rightly be inferred that the techniques used for fossil fuels may not be optimal in the search for geothermal energy. It has already been emphasised that, although the crust of the Earth is imbued with a tremendous heat reserve, diffusion of this which generally occurs is the reason why geothermal energy pockets with higher than average heat contents have to be found for exploitation. The actual grade of any such resource has to be stipulated according to the uses envisaged and will vary. Development will also proceed at a pace dictated not only by economics, but also by technology. Our present drilling activities in the geothermal energy field do not go very deep and there is no doubt that improving technology will facilitate descending further than say the deepest well at Larderello (about 2·7 km[1]), perhaps to depths as great as 10 km or even more. Jim Combs and L. J. P. Muffler believed that naturally productive geothermal systems require the circulation of water, but hot, dry rock can be *made* productive if water can be introduced into it through fractures, artificial or geological.[2] In other words, porosity and permeability values must be adequate for a circulation pattern to be set up. Where these are inadequate, artificial stimulation by explosions, thermal cracking and other methods may solve the problem.

3.1. INTRODUCTION

It must be stated at the outset that exploration techniques are still in a developmental stage and have mostly relied on geophysical investigations up to now. Additionally, they are usually also confined to regions having surface expressions such as hot springs and fumaroles. However, geochemical researches are becoming ever more important and recently there have been important advances which will be looked at below.

To begin with, the geological factors ought to be examined. After all, the problem is basically one pertaining to the earth sciences.

It has been seen that geothermal reservoirs are not homogeneously distributed in the planetary crust, but rather occur near the margins of crustal plates,[3] specifically near those margins where crust is in the process of being consumed or created, i.e. where molten rock is being generated and is in the mobile state moving within the crust. Intrusion thereafter provides heat; subsequent transfer to convecting systems of meteoric water results from conduction. Of course, tectonic activism is the chief characteristic of such margins and manifests itself in the form of volcanism and anomalously high heat flows of either a regional or localised nature. Sedimentary basins in appropriate locations are involved, but more interesting is the case of those far removed from plate margins which nevertheless contain hot water often deeper than 3 km and at pressures exceeding hydrostatic. A case in point is the Hungarian resource of geothermal energy, but others are known in the Gulf Coast and also in the USSR.[4,5,6] Here, a combination of lowered thermal conductivity and higher heat flow may be responsible. Shield areas are useless, being tectonically inert, devoid of recent volcanism and composed of very old metamorphics and low-permeability igneous rocks. Shields comprise the cores of many continents, for instance the Canadian Shield, the Scandinavian Shield, etc.

The heat source below a geothermal system (Fig. 3.1) is most likely to be a deep-lying igneous mass more than 5 km down and this will be the driving

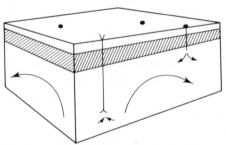

FIG. 3.1. A geothermal reservoir model, well on the right being an injection type, that on the left an extraction type. Circulation patterns in ground-water movement are shown and the overlying hachured stratum represents a semi-permeable or impermeable cap rock.

force behind the overlying meteoric convection system.[7] A concomitant of this is that the overlying rock units just happen to be there and could possess the necessary physical characteristics of porosity and permeability through a range of lithological properties. Also important are the structural features of these rocks. Obviously, structurally favourable parameters such as joints and fracturation will favour development of a geothermal reservoir because hot water will rise along the path of least resistance, not necessarily vertically but any way upwards. Geometric considerations may display the rising water and it may eventually emerge kilometres away from the subterranean source. Hot-water convection systems will have an influence on the actual rocks as well. For example, silica will go into solution at depth and may be reprecipitated on the ascent as quartz or as opal or as beta-cristobalite, thus sealing off the periphery of the rising hydrothermal plume.

3.2. METHODS OF EXPLORATION

There are a number of methods of exploration, some geophysical and geochemical, others not. All have the same objective, however, and that is the finding of and delimiting of regions which are underlain by hot rock. It is also very necessary, if possible, to make a reasonable estimate of the volume involved as well as the permeability and temperature below. A prediction of the chemical composition of the fluids is valuable, where feasible. The varieties of geothermal system have already been indicated in Chapter 2 and the differences between these correctly implies the fact that many exploration methods are required for the various circumstances which may arise. Some possible approaches may now be outlined.

3.2.1. Search for Information

The search for information is almost always useful because, in most areas, previous data regarding the topography, geology, hydrology, meteorology, geochemistry and geophysics exist, can be compiled and furnish valuable background briefing. Satellites also, by remote sensing, can give a lot of extra information. Where such data can be obtained, much effort is saved on the survey side and also facts are available which can be utilised in interpreting additional data derived from the actual geothermal survey.

In a few cases, prior information is very scanty, however, and the writer encountered this in investigating the area of Maku–Khoy in Iran about which very little is known. The Imperial Iranian Geological Survey has published a 1:1 000 000 geological map of the northwest of that country which includes the suspected geothermal area and from it, it is clear that Maku–Khoy is a highly tectonised and fracturated, primarily igneous zone with evidence of extensive and fairly recent volcanism. It is most unfortunate that access is very difficult, roads being inadequate and the topography, itself at an uncomfortable elevation, very rugged. As a result, a lot of ground work cannot be performed, although occasionally helicopters may be employed. An additional hazard here is the proximity of two international frontiers, namely the Turkish on the western periphery and that of the USSR to the north. In 1978, while the writer was engaged in examination of this part of Iran, a helicopter crash with fatalities occurred.

3.2.2. Aerial Surveys
Aerial photographs, if properly obtained, can be invaluable—if they are correctly interpreted, a specialist job. Until this method became available, geologists were compelled to rely upon ground observations and topographic maps alone and, as at Maku–Khoy, this would have meant very scanty results. Photo-interpretation has now become one of the most potent geologic methods in such areas. Map making from aerial photographs is a part of photogrammetry and the geologist's skill in interpreting land forms is critical. Sometimes the method may be supplemented by the SLAR (side-looking airborne radar) technique which has one advantage (v. Appendix 2). It can be done at any time whereas aerial photographs must be made at suitable hours when shadow effects are just right. The normal practice with these latter is to use pairs of adjacent photographs stereoscopically, marking geological information either on the prints or on a sheet of transparent material put over them. One may estimate the dips of beds exposed on dip-slopes or in gorges, but with visual inspection, it is to be remembered that the stereoscope enlarges the vertical scale so that both the dips and the topography are exaggerated. Actually, photographs reveal most in arid or semi-arid regions, so are particularly useful in the Middle East. However, useful results can be obtained in even densely vegetated and tropical terrains, the vegetation sometimes reflecting the topography. For instance, in New Guinea, the practically constant heights of tree tops above the ground surface allows the geologist to recognise escarpments reproduced in the foliage! It has been noted that previous experience in the field in any area facilitates the correct appraisal of aerial photographs.

In cases where the regional stratigraphy and structure are not well known, an aeromagnetic survey is sometimes valuable. For detecting surface thermal manifestations, an airborne infrared survey is very useful.

However, neither of these methods will yield direct information regarding the subterranean distribution and type of geothermal resource in a region.

3.2.3. Geological and Hydrological Survey

A geological/hydrological survey may be effected in order to delimit the thermal area in a geographical sense. The geological component of the survey will investigate particularly the tectonics and stratigraphy of a region. Very interesting and relevant characteristics are recent faulting and volcanism as well as, of course, thermal manifestations such as hot springs, fumaroles, hydrothermally altered rocks, etc. The hydrological component should take in temperature and discharge measurements on hot and cold springs and chemical analyses of them. The water-table levels should also be noted from available wells. If possible, variations in this over the year should be determined as well. The surface and subterranean movements of water must be studied. Meteorological information of value included precipitation and humidity, factors important in subsequent planning operations preceding drilling.

Once a geothermal prospect is identified as a result of the above-mentioned reconnaissance activities, then detailed studies are effected around the locations of potential drilling with the object of forecasting the conditions which are likely to be encountered at depth. Active faults are often selected as drilling targets.[8] It is well known that the type and character of vegetation vary with the temperature of the soil and can be used, therefore, in order to delimit thermal zones.

From the observations made in the geological and hydrological survey, it is possible to construct a model incorporating stratigraphical, petrological, hydrological and structural information for use as an indicator of the correct direction to take in later exploration and exploitation work.

3.2.4. Geochemical Survey

A geochemical survey results from sampling of waters and gases from hot springs and fumaroles and also from subterranean sampling from boreholes and wells and the analyses of these enable the determination of whether the geothermal system is, for instance, hot-water or vapour-dominated to be made. Additionally, it is possible to estimate minimal temperatures to be anticipated at depth, assess the homogeneity of water supply and find out the chemical character of water at depth. Also, the source of recharge water can be determined.

In recent years, some startling advances have been made and it is proposed to discuss some of these in detail.

3.2.4(i) Chemical Geothermometry
Nowadays, qualitative and quantitative chemical geothermometers are
employed in exploration for geothermal energy as routine tools. The latter
require chemical analyses of thermal waters from springs or wells.
However, qualitative techniques can be employed in searching for
anomalous concentrations of appropriate indicator elements in a wide
spectrum of sources such as soils, soil gases, fumaroles, hot and cold springs
and even rivers.[9]
   The assumption is made that the relevant indicator dispersed from a heat
source at depth. A possibility still awaiting investigation is that soil and gas
analyses may well present the chance of detection of hidden geothermal
sources devoid of surface manifestations (an entirely different approach
discussed below in Chapter 11 derives from the fact that high seismic noise
amplitudes have been recorded in geothermal areas such as the Yellowstone
National Park). It is important to discuss the two categories of chemical
geothermometers alluded to above.[9] Other, less well-established ones exist
and two of these which are still, to a degree, at the experimental stage will be
described in Chapter 11 also; these are the CO_2–methane and the
CO_2–H_2O geothermometers.

1. Qualitative geothermometers
   Most of these are based upon the distribution and relative concentration
of volatile elements in soils and waters or variation in the composition of
soil gas. Around shallow heat sources, anomalously high concentrations of
volatiles may accumulate, especially if subterranean boiling takes place.
Similar anomalous concentrations may be found in situations where
favourable structures permit the escape of such volatiles from deep crustal
regions where metamorphic reactions are going on. What Fournier termed
fossil hot springs (i.e. cold springs) and deposits of ores may exhibit halos of
volatiles which, in practice, are difficult to distinguish from halos resulting
from contemporary activity.[9] F. Tonani has suggested that enrichment of
B, NH_4, HCO_3, Hg and H_2S in near-surficial waters may result from
boiling at depth.[10] Volatiles are partitioned into the steam phase and
ascend eventually to become incorporated into the shallow and relatively
cold ground water. These are probably the same volatiles most
characteristic of vapour-dominated systems so that their concentration at
high level in such shallow ground waters may well indicate that such a
system is present. W. A. J. Mahon has noted that high Cl/F and Cl/SO_4
ratios in liquids from a geothermal area indicate high temperature in the
system and also that variations in CO_2/NH_3, CO_2/H_2 and CO_2/H_2S from

fumaroles may indicate zones probably proximate to deep hot water.[11] He stated that when both high and low ratios of these gases occur, fumaroles with the *lowest* ratios may be the closest to the aquifer. G. E. Sigvaldson and G. Cuellar believed that hydrogen in thermal gases may be regarded usually as a qualitative index of high temperature and that quantities exceeding 0.5% mean reservoir temperatures in excess of 200 °C.[12] R. O. Fournier and A. H. Truesdell stated that the mole ratio $Cl/(HCO_3 + CO_3)$ is valuable in distinguishing waters from different aquifers in Yellowstone National Park, higher ratios defining waters originating in hotter aquifers.[13] The association of mercury and thermal springs in California was noted as long ago as 1935 and M. Dall'Aglio, R. Da Roit, C. Orlandi and F. Tonani demonstrated that the mercury content of stream sediments defines a clear halo round the Larderello and Monte Amiata Italian geothermal regions.[14] Prospection using mercury anomalies is facilitated by the recent development of highly sensitive and rapid detection methods which are applicable to soils and soil gases.[15] Even more recently, R. W. Klusman and his associates completed an evaluation of the mercury concentration in soils of six Colorado geothermal areas utilising regression analysis. This incorporated the secondary effects of pH and the organic carbon, iron and manganese concentrations on mercury in soils. The technique was utilised in order to evaluate a geothermal anomaly at Glenwood Springs, Colorado. Results indicate that secondary influences on mercury in soils of geothermal areas may be of importance in evaluating areas of anomalous mercury leakage.[16]

It is quite possible that anomalously high concentrations of non-volatiles also may indicate high subsurface temperatures.[9] This would depend upon the capacity of the element to be leached from rock at high temperatures and not leached at low ones. It has been reported, for instance, that high lithium concentrations in Tuscan springs and streams provide a good indication of high subsurface temperatures at relatively shallow depths.[17] Unfortunately, the method is not foolproof because variations in trace-metal content can arise from factors other than high subsurface temperatures. These include lithological variations and variations in the time during which the water was in contact with rock and also contamination.

A very new approach has now appeared and that is the use of helium variations in soil gas.[18] Good correlation has been reported between temperature gradients and observed helium concentrations over hidden geothermal prospects at the Dunes, East Mesa and Brawley geothermal areas of the Imperial Valley in California.

2. Quantitative geothermometers

In the utilisation of chemical compositions of spring and well waters in order to determine as well as possible, really to guesstimate, subsurface temperatures, a number of assumptions have to be made and these are as follows:

(a) temperature-dependent reactions involving rock and water determine the amount or amounts of dissolved indicator constituents in the water;

(b) there is a sufficient supply of all the various reactants;

(c) there is equilibrium in the aquifer or reservoir in respect of the specific indicator reaction;

(d) no re-equilibration of the indicator components takes place after the water leaves the reservoir;

(e) either there is no mixing of different waters during ascent to the surface or it is possible to evaluate the consequences of such mixing if it takes place.

Obviously, the attainment of equilibrium in the reservoir depends upon several factors, for instance the temperature and the reactivity of the country rock as well as the kinetics of the particular reaction. Clearly,. therefore, equilibrium cannot always be guaranteed.

Two kinds of temperature-dependent reactions which may prove to be of use as quantitative geothermometers are solubility and exchange reactions.

Solubilities: Mineral solubilities alter as functions of both temperature and pressure. Curiously, only silica has been extensively employed as a geothermometer. In the majority of natural waters, dissolved silica is uninfluenced by common ion effects, the formation of complexes and loss of volatile components, factors which make interpretation hard for most other dissolved constituents. Significantly, the assumption of an adequate supply of silica is justifiable and this is certainly not the case for other reactants. R. O. Fournier gave an excellent instance of this when he noted that a geothermometer based upon the solubility of fluorite (CaF_2) is useless in the event of the absence of this mineral where water–rock chemical equilibration occurs.[9] Actually, Fournier and J. J. Rowe established a silica geothermometer based on quartz solubility and utilised it to estimate subsurface temperatures in hot-spring systems.[19] They gave curves by the use of which the silica content of water from a hot spring or well can be correlated with the last temperature of equilibration with quartz and discovered that the quartz geothermometer is best suited to the temperature range 150–225 °C. At higher temperatures, silica may well deposit during the rising of the water. At lower temperatures, other silica species such as chalcedony, cristobalite or amorphous silica may control dissolved silica. Equations relating the solubility C (in mg SiO_2/kg water) to temperature t in the range 0–250 °C of various pure silica minerals are as follows:[9]

Amorphous silica:

$$t°C = \frac{731}{4·52 - \log C} - 273·15$$

Beta-cristobalite:

$$t°C = \frac{781}{4·51 - \log C} - 273·15$$

Alpha-cristobalite:

$$t°C = \frac{1000}{4·78 - \log C} - 273·15$$

Chalcedony:

$$t°C = \frac{1032}{4·69 - \log C} - 273·15$$

Quartz:

$$t°C = \frac{1309}{5·19 - \log C} - 273·15$$

Quartz (after steam loss):

$$t°C = \frac{1522}{5·75 - \log C} - 273·15$$

H. Sakai and O. Matsubaya employed experimental data of W. L. Marshall and R. Slusher giving the solubility product of anhydrite from 100 °C to 200 °C at various ionic strengths and utilised the concentration product of Ca^{2+} and SO_4^{2-} in order to estimate subsurface temperatures in hot-spring systems. The estimated temperatures were in accord with temperatures estimated from the oxygen isotope fractionation between sulphate and water.[20,21] This anhydrite geothermometer is expected to be most useful in thermal waters with rather low concentrations of Ca^{2+} and HCO_3^-, i.e. waters in which the Ca^{2+} concentration is not controlled by the solution and precipitation of carbonate.

Exchange reactions: The equilibrium constants for exchange reactions are, of course, temperature-dependent and in such reactions, the ratios of dissolved constituents alter with changing temperatures of equilibration. Many possible constituents and reactions are potentially useful. Some instances are

(a)   Na/K ratios of alkali chloride solutions equilibrated with alkali feldspars;[22]
(b)   Na/K ratios in natural waters;[13]
(c)   Na–K–Ca relations in natural waters.[23]

The Na/K ratio has been found to work well for the estimation of temperatures of waters above 200 °C, but where the waters are below 100 °C, anomalously high estimated temperatures are obtained. For such low-temperature waters, the Na–K–Ca geothermometer has proved to be more reliable. The equation is

$$\log(\text{Na/K}) + \beta \log(\sqrt{\text{Ca}}/\text{Na}) = \frac{1647}{273 + t\,^{\circ}\text{C}} - 2\cdot24$$

where concentration units are mol/kg and $\beta = 1/3$ for water equilibrated above 100 °C and 4/3 for water equilibrated below 100 °C. T. Paĉes recommended application of a correction factor for waters below 75 °C, with partial pressure of CO_2 in the aquifer above 10^{-4} atm.[24] The correction factor, $I = -1\cdot36 - 0\cdot253 \log P_{CO_2}$, is subtracted from the right-hand side of the above equation.

The ratios $\sqrt{\text{Ca}}/\text{K}$ and $\sqrt{\text{Ca}}/\text{Na}$ may perhaps be employed for the estimation of temperatures.[23] There is some evidence that springs more directly supplied from hot aquifers possess the highest Na/Ca ratios. If this is indeed so, it may reflect the retrograde solubility of calcium carbonate.

3.2.4(ii) Rising Waters
Obviously, hot waters will cool during ascent to the surface and this may take place in a number of ways. For instance, it may be adiabatic by boiling or it may be effected by conduction. Another possibility is that of mixing with shallow and cooler waters. To a degree, the chemical analyses of spring waters provide useful information as to which of the above processes is actually taking place—or perhaps, which combination. From the geochemical standpoint, those waters which rise both directly and rapidly from aquifers with a minimal amount of conductive cooling are the simplest

to investigate because their composition will mirror the rock–water equilibrium at the aquifer temperature. If this is below atmospheric boiling, water is able to emerge at roughly the temperature of the aquifer. In the case where the aquifer is at a higher temperature than the atmospheric boiling temperature, the water will cool in an adiabatic manner and emerge as a boiling spring, steam perhaps being separated during the ascent. If this is slow, then the possibility or even probability of conductive cooling *en route* must be reckoned with. The consequence is that, should such waters move laterally and emerge in a series of springs, different discharges may have different temperatures but will retain the same chemical composition. The most complex situation exists where hot waters rise and mix with cooler overlying ground water because these two waters will have different chemical compositions. Attempts must then be made to set up mixing models. Some observations on these may be made.

Mixing models

Clearly, in nature, there must be many cases of hot springs emitting waters which represent mixtures of the type alluded to above and quite possibly this is generally the case. Unfortunately, such mixtures probably rarely attain chemical equilibration and even if chemical equilibrium is reached, chemical geothermometry will only indicate the temperatures of the *mixed* waters and not that of the *hot*-water component. It is apparent from this that quite independently of whether chemical equilibrium is established after mixing occurs, the hot-water component temperature is impossible to estimate from a solubility relationship *unless* the mixing is taken into account. However, such a hot-water component temperature estimate may be almost unaffected by the mixing process if based upon an exchange reaction using a ratio of dissolved constituents provided that

1. the indicator elements are much more concentrated in the hot water and relatively depleted in the shallow, cool water, and
2. the relative concentrations in the two components of mixing stay more or less the same after mixing takes place.

Obviously, dilution will affect the Na–K–Ca geothermometer because the calculation involves the square-root of the concentration.

R. O. Fournier and A. H. Truesdell have described two mixing models which may be applied to springs having large flow rates and temperatures below boiling.[25] In using them, the silica content and temperature of the hot spring and the cold water must be determined.

Model 1: Here, the enthalpy of the hot water and steam mixing with and heating the cold water is the same as the initial enthalpy of the deep, hot water, i.e. the latter may boil before mixing, but all the steam condenses into the cold water.

Model 2: In this case, the enthalpy of the hot water in the zone of mixing is less than that of the hot water at depth because of the escape of steam during the rising process.

In both models, it is requisite that the initial silica content of the deep, hot water be controlled by the solubility of quartz, and that no further solution or deposition of silica takes place prior to or after mixing. Truesdell and Fournier have devised a simple procedure for application of both models and this uses a plot of dissolved silica against enthalpy.[26]

Neither model is applicable in the case of boiling springs as a result of the fact that there is a loss of heat in the steam after mixing. Here, another mixing model is necessary in order to calculate subsurface temperatures deeper in the system than would be obtainable from the normal method of applying the silica or quartz geothermometer. In practice, the plot of enthalpy against chloride is employed and it functions optimally where the initial temperature of the hot-water component exceeds 200 °C. Again, some assumptions are made and these are

1. that notable heat losses or gains do not take place prior to or subsequent to mixing;
2. that re-equilibration with quartz takes place after mixing;
3. that silica is not precipitated during the rising of the mixed water to the point of sampling on the surface.

Where spring waters cool mainly by conduction, a somewhat higher estimated subsurface temperature and enthalpy would be derived by using the silica geothermometer and the chloride content of the deep water would be the same as that of the emerging spring water.

3.2.4(iii) Isotopic Techniques in Geothermal Investigations
The chemical composition of water is useful in this matter, comprising as it does hydrogen, deuterium, tritium, oxygen-16 and oxygen-18.

These stable and radioisotopes may be employed in identifying sources and ages as well as mixing processes and trends in time.

The water molecule contains one atom of oxygen and two atoms of hydrogen. Since tritium content is extremely small, the normal water species are $H_2{}^{16}O$, HDO (abundance 320×10^{-6}) and $H_2{}^{18}O$ (abundance 2×10^{-3}), the actual concentrations of the stable isotope forms in terrestrial waters undergoing small variations of up to 30% for HDO and 5% for oxygen-18 water. Both deuterium and oxygen-18 may be used as natural tracers. B. Arnason has measured deuterium in hot water in Iceland and thereby deduced its origin and indeed the flow paths of most of the hot, ground-water systems in the country.[27] He was able to demonstrate that the thermal waters are of meteoric origin, being derived from precipitation on the mountainous interior. This descends to great depth and becomes heated as a consequence of the geothermal gradient. Thereafter, flow takes place into topographically lower regions and escape is effected through fissures and faults of which there are many and also beneath the ocean floor. The utility of the approach is based upon fractionation phenomena in nature whereby, for instance, precipitation becomes isotopically lighter, i.e. depleted with respect to heavy isotopic species, with increasing latitude and altitude and onset of cold seasons. It has also been used elsewhere, of course, for example in Japan by H. Sakai and O. Matsubaya.[28] Studies were made on Arima-type brines and Green-Tuff-type thermal waters. Three volcanic systems were involved, namely Hakone (a subaerial volcano), Ibusuki (a half-drowned caldera) and Satsuma-Iwojima (a volcanic island).

Another application may be cited and this was made by W. F. McKenzie and A. H. Truesdell who estimated geothermal reservoir temperatures from the oxygen isotopic compositions of dissolved sulphate and water from hot springs and shallow drillholes.[29] Three areas in the western USA were used and the limited analyses performed suggest that dissolved sulphate and water are most probably in isotopic equilibrium in all reservoirs of significant size, also that little re-equilibration occurs *en route* to the surface.

H. Sakai has examined sulphate–water isotopic geothermometry as applied to geothermal systems.[30] The SO_4–H_2O oxygen isotope thermometer utilises the fact that the δ^{18} oxygen values of SO_4^{2-} in geothermal waters are determined by the isotopic exchange reaction

$$\tfrac{1}{4}S^{16}O_4^{2-} + H_2{}^{18}O = \tfrac{1}{4}S^{18}O_4^{2-} + H_2{}^{16}O$$

and the isotope separation factor

$$\alpha = (^{18}O/^{16}O) \text{ in } SO_4^{2-}/(^{18}O/^{16}O) \text{ in water}$$

is a function of temperature.

The rate of the oxygen isotopic exchange reaction between sulphate and water in acidic to neutral thermal waters of a temperature of 100 °C or more is fast enough to justify the statement that the sulphates are in isotopic equilibrium with the reservoir waters. Successful applications of the approach have been made to several Japanese fields.

As regards the use of isotopic technology in attempting to date ground water, this employs two radioisotopes, tritium and radiocarbon. It is with the former that we are concerned here since it can be a component of the water molecule, even if only in infinitesimal quantities.

Tritium

This is the radioactive isotope of hydrogen and it emits beta-radiation (maximal energy 18 keV), having a half-life of 12·26 years. It occurs in precipitation and therefore infiltrates to subterranean waters as a result of three processes which are as follows:

1.   stratospheric production as a consequence of the nuclear reaction between cosmic-ray-generated neutrons and nitrogen, thus

$$(^{14}N, n)(^{3}H, ^{12}C)$$

2.   solar emission,
3.   thermonuclear detonations (since 1952).

Amounts due to (1) and (2) are about 10 tritium units (TU), one TU being one atom of tritium per 10^{18} atoms of ^{1}H. The amount due to (3) was once greater by several orders of magnitude, even reaching 10 000 TU. Due to the thermonuclear moratorium, however, decline occurred, but atmospheric levels still exceed natural ones.

Tritium is an indicator of recent recharge in subsurface waters—*v.*, for example, the present writer's studies with P. W. Williams.[31,32] A water sample containing significant quantities of it can be considered as containing at least a component of modern recharge if it is assumed that no contamination has occurred after it left the aquifer.

The principle of radioactive decay is theoretically applicable in order to determine the time which has elapsed since the sample entered the ground-water system. The relation may be expressed

$$C = C_0 \exp(-\lambda t)$$

where C is the concentration of tritium at some observation point in the aquifer at a time t after recharge, C_0 is the concentration of tritium at the point of recharge and λ is the decay constant for the radioisotope ($\frac{1}{18}$). Needless to say, assumptions have to be made which are difficult to justify, for instance that C and C_0 are constant with time, so that C/C_0 is related to the transit time in the aquifer. The problem here stems from the fact that very little is known about the values of C prior to the 1950s. Also, the model implied in the above relation is a piston-flow one without dispersion or mixing, both of which frequently occur. By refinements, however, tritium can be applied to the determination of residence time in well-mixed reservoirs and its decay can be related to the velocity of ground-water movement.

The dating span of tritium extends over a half century or so (four half-lifes) and if much older waters are involved, radiocarbon is the appropriate tool so that a few words regarding it might be apposite.

Radiocarbon

The principle of its employment is the same as for tritium, but, in view of its much greater half-life, 5730 years, its range is larger, extending over 25 000 years. Chemically, carbon-14 is also a beta-emitting radioisotope (maximal energy 156 keV) produced in the upper atmosphere as a result of the interaction of cosmic-ray-produced neutrons and nitrogen, thus

$$(^{14}N, n)(^{14}C, {}^{1}H)$$

The carbon is oxidised to carbon dioxide which then mixes into the CO_2 reservoir, the content in water remaining constant until this is removed into the ground-water system when there occurs a diminution governed by the law of radioactive decay. This is the basis of its utilisation in determination of age and transit time of ground-water samples.

3.2.4(iv) Radiogenic Noble Gases in Geothermal Tracing
One of the most significant gases involved is radon, ^{222}Rn, a noble gas formed by the radioactive decay of ^{226}Ra and having a half-life of 3·8 days, therefore being brought to the atmosphere only over continents. The half life of ^{226}Ra is 1620 years and this element is uniformly distributed in the crust at an average concentration of 1 pg/g or thereabouts.[33] Radium undergoes geochemical processes such as adsorption, co-precipitation in hydrothermal systems, etc.

The rate of release of radon from the ground is not uniform, depending on rock type, soil type and the physical nature of the soil (water content being significant) as well as changes in atmospheric pressure. Air masses over oceans can be almost radon-free if they have not been over land for weeks, but air masses over continents can become radon-enriched up to a point where the radioactive decay of the nuclide balances the rate of its release from the ground. Since this latter phenomenon takes up to a few half-lifes, the radon content of air can be a measure of the time the air masses have spent over continental regions. Like Rn, atmospheric neon, argon, krypton and xenon tag waters by solution. Where ground water is heated, i.e. in a geothermal system, the gases pass into the steam phase and thus deplete the residual waters. Radiogenic He, Ar and Rn enter deep-seated hot waters by flushing from country rocks in which they are formed.[34]

As the noble gases are inert, they furnish us with excellent conservative tracers which can provide information on a variety of matters such as

1. indications of meteoric origin,
2. degree of mixing between shallow and deep waters,
3. previous heating episodes,
4. mechanism of formation of subterranean steam (this may possibly be derivable by studying the degree of fractionation between light and heavy noble gases),
5. interconnections between adjacent producing wells and therefore the forecasting of optimal spacing in a steam field.

3.2.4(v) The Chemistry of Geothermal Gases
This question has been studied by, among others, F. D'Amore and S. Nuti.[35] They looked into some chemical reactions which may well justify the presence of the more common gas species and their relative abundance in geothermal fluids. Among these are H_2O, H_2S, NH_3, N_2, CO_2 and methane. The major problem is to reconstruct deep, thermodynamic conditions from analytic data derived from surface manifestations or wells and the authors named used information from the examination of Larderello. A summary of results may be useful.

1. Hydrogen content and oxygen fugacity: The water dissociation reaction is ubiquitous because of the very wide range of conditions under which water is the dominant component and also because of the fact that water has a very long residence time. The reaction is expressed

$$H_2O \underset{\longleftarrow}{\overset{K}{\longrightarrow}} H_2 + \tfrac{1}{2}O_2$$

K, the equilibrium constant, varying with temperature and

$$\log K = 2.94 - \frac{12.81 \times 1000}{T}$$

where T is the temperature in K.

2. H_2S content and sulphur fugacity: Sulphur fugacity (S_2 gas) can be buffered locally in reducing conditions by the pyrite–pyrrhotite equilibrium reaction expressed thus:

$$2FeS_2 = 2FeS + S_2 \text{ (gas)}$$

Pyrite has been confirmed in the Larderello basement and D'Amore and Nuti consider it to be in equilibrium with pyrrhotite in hydrothermal conditions.

3. NH_3/N_2 contents: Fluids at Larderello are ammonia-rich, probably because of thermal degradation of the nitrogen-rich organic matter in the underlying Palaeozoic–Triassic sedimentary layers. D'Amore and Nuti attempted to show that a few parts per million of dissolved ammonia in the deep geothermal water would give NH_3 contents at the surface very like those actually observed in the produced steam.

4. CO_2/CH_4 contents: These gases are apparently always indicative of a carbon source which may be either an environment containing excess carbon dioxide or one which contains elemental carbon as, say, graphite or in some other form, or carbon bound in bituminous substances of asphaltic nature or unsaturated hydrocarbons.

Many natural sources will be organic, of course, but the most important ones are no doubt represented by the large group of reactions affecting rocks which contain carbonates and various forms of silica (e.g. quartz, clay minerals).

3.2.5. Geophysical Survey
Geophysics is utilised mainly in the defining of target areas for actual drilling after a geothermal prospect has been identified broadly. The measurement of various physical parameters at depth, for instance temperature, electrical conductivity, the propagation velocity of elastic waves, density and magnetic susceptibility, can confirm the existence of a geothermal reservoir and help to delimit it. It has been found that the most

60   GEOTHERMAL RESOURCES

useful geophysical approach for geothermal exploration is that of assessing temperature or the geothermal gradient survey, determining heat flows, conducting electrical-conductivity investigations and using seismic methods.

3.2.5(i) Thermal Methods

It has been seen earlier that geothermal areas have anomalously high heat flows (although these can sometimes arise from other causes such as exothermic chemical reactions or the concentration locally of radioactive minerals). Those in prospect regions are quite restricted in geographical extent and last for a very long time, namely for millennia. Elevation of heat flow may exceed normal by orders of magnitude if a body of magma or rocks of near-magmatic temperature are nearby. The techniques of thermal exploration are invaluable in estimation of the actual size and therefore the potential of a geothermal system and comprise

1. surficial and shallow temperature measurements. By shallow is meant depths of up to 6 m or so,
2. geothermal gradient surveys ranging up to 100 m in depth,
3. a more expensive approach, namely heat-flow determinations at depths in excess of 100 m.

The difficulty with the cheaper surficial and shallow measurements is that they are greatly influenced by near-surficial effects such as precipitation, infiltration and movements of ground water as well as topographical factors. Many investigators have utilised temperature measurements as a function of depth in boreholes anything from 15 m to 150 m deep as a means of geothermal exploration, optimal depths in many ways because they are free of near-surface thermal disturbances. Also, gradient measurements can be made here with a high degree of precision. However, the possible effects of lateral movements of ground water must be borne in mind. It is a fact that in most geothermal areas which offer commercial and economic possibilities, the gradients at such depths are over 7 °C per 100 m. Normal geothermal gradients are less, more of the order of 2–3 °C per 100 m. One of the difficulties arises from the danger of extrapolating measured gradients to greater depths. If this is done, too high values will probably result. The reason is two-fold. First, there is usually great variation in conductivity of rock with depth. This may be due to a number of factors, one of which is that porosity diminishes downwards. Minerals are much more conductive than water, in fact up to ten times more. Hence, bulk conductivity increases greatly with decreasing porosity (increasing depth). Heat flow is the product of gradient and thermal conductivity; therefore, as thermal conductivity increases with depth, the thermal gradient must decrease with depth. Secondly, convection will have a very significant effect in reducing thermal gradients at depth. Actually, convection is thought to produce high conductive gradients in rock above a convection cell, but very low thermal gradients within the convection cell itself.[36] In regions where water convection *within* pore spaces of rocks is feasible, extrapolation of measured high near-surficial gradients is not justifiable and may very well yield much too high temperatures at depth.

Sometimes, the mean thermal conductivity of the subterranean strata may remain practically constant throughout the zone of drilling of boreholes and in this case, it is clear that measured thermal gradients are proportional to heat-flow values. However, the measurement of heat flow is more useful than is that of thermal gradients in that the former is quite independent of the thermal conductivities of the various rock types. Consequently, where non-homogeneity is encountered, they alone afford accurate information as regards the potential geothermal energy production area.

It is apparent that, ideally, both types of measurement, i.e. both gradient and heat flow, should be used and, in fact, it has been noted that the latter alone is valuable in defining broad areas of anomalously high heat flux. Such regional heat flows exceed normal and where this is coupled with heat-generation measurements from bedrock samples which are normal, the presence of hydrothermal convective systems and also perhaps of recent and hot intrusives may well be inferable.

D. T. Hodder and others utilised the radiation emitted primarily in the infrared region as a method of mapping thermal activity.[37] The operation of the infrared scanners is usually in the 3–5 micron or 8–14 micron transmission windows in the atmosphere. However, thermal-infrared imagery appears to be a noise-limited system. Many of the 'thermal' anomalies on the imagery may arise from outcrop, slope direction and degree, moisture content of soil, humidity, variations in the properties of rocks, etc. Actually, it is difficult at present to detect heat-flux anomalies lower than at least 150 times normal.

3.2.5(ii) Electrical and Electromagnetic Methods
These measure electrical conductivity at depth and depend upon the fact that factors such as temperature, porosity, interstitial fluid salinity, etc., tend to be higher in geothermal reservoirs than in surrounding ground.

Thus, the electrical conductivity will also be higher. A number of approaches can be made to measure electrical resistivity at depth. This is inversely proportional to the above-mentioned parameters and also to permeability, another factor which must be adequate to enable a geothermal reservoir to exist. Telluric and magnetotelluric methods which depend upon measuring variation in natural electrical or electromagnetic fields therefore measure electrical resistivity at depth. The resistivities of selected materials are cited in Table 3.1.[38]

TABLE 3.1

RESISTIVITIES OF SELECTED MATERIALS

Material	Electrical resistivity (ohm-metres)
Copper	10^{-8}
Pyrite	10^{-6}
Concentrated salt water	2×10^{-2}
Clay, wet and plastic	1–3
Lignite or gypsum	10^3
Oil	10^4
Calcite or dense limestone	10^7
Quartz	10^{10}

Clearly, in the above table, quartz is the worst conductor, copper the best. Electrical-resistivity testing may utilise several approaches, for instance horizontal exploration, vertical soundings, and resistivity *minima* indicate the existence of geothermal reservoirs. Archie's equation is applied:

$$\rho_o/\rho_w = \phi^{-m}$$

where ρ_o is the formation resistivity, ρ_w is the water resistivity obtained directly, ϕ is the porosity and m is an estimated cementation factor. This will provide means of calculating the resistivities of a number of formations which can thereafter be compared in order better to define the geothermal reservoir (Fig. 3.2). As well as direct current methods, induction electromagnetic methods have also proved very useful. Methods of electromagnetic induction prospection include

1. audio-frequency magnetotelluric method for depths up to a few hundred metres;

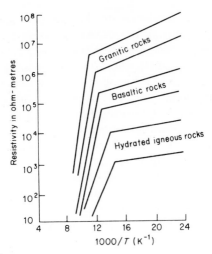

FIG. 3.2. Resistivity of various types of dried rocks (granitic, basaltic and hydrated igneous) as a function of temperature. From J. J. Jacobson (1969), *Deep electromagnetic sounding technique*, D.Sc. Thesis T-1252, Colorado School of Mines.

2. two-coil profiling method providing high resolution at depths from 20 m to 30 m;

3. two-coil frequency-domain sounding method for depths from 1 km to 2 km;

4. coil/wire time-domain sounding method which may be used in order to map the source of anomalous heat into the lower crust and upper mantle.

Abhijit Dey and H. F. Morrison have provided an analysis of the bipole–dipole method of resistivity surveying, a method widely used in reconnaissance work for geothermal exploration. Apparent resistivities are plotted at roving dipole receiver locations and the current source (bipole) is fixed.[39] Usually, interpretation is in terms of simple, layered, dyke, vertical contact or sphere models. Ambiguity arises in the case of more complicated two-dimensional models. Here, the detection of buried conductors depends upon the choice of transmitter location. Dey and Morrison compared the two methods bearing in mind that, since apparent resistivities taken on a line collinear with the dipole are roughly equivalent to the apparent resistivities for one sounding in a dipole–dipole pseudo-section, this approach is permissible.

Magnetic methods depend upon the fact that changes occur in the magnetic properties of magnetic minerals. Those of volcanic rocks are variable and depend upon original lava flow and the rate of cooling, the homogeneity of the material, the surface topography and other factors. Hydrothermal alteration and temperature changes also affect magnetic properties. For instance, the former will convert magnetic magnetite into non-magnetic pyrite. If the temperatures exceed the Curie point, then magnetic minerals will lose their magnetic properties altogether. Of course, this is not considered to be a likely occurrence because the Curie point of the majority of magnetic minerals far exceeds temperatures found in geothermal reservoirs. Use of a magnetometer enables location of magnetic anomalies to be made and from these the magnetic contours of an area can be mapped. Magnetic surveys can facilitate the determination of subsurface rock types, the occurrence of hydrothermal alteration and perhaps even find deposits of hot rocks.

3.2.5(iii) Gravimetry
In the right circumstances, gravity surveys can delineate major structural features and pinpoint local positive anomalies which may relate to the geothermal system. Such positive anomalies can be caused by local structural highs, buried volcanic rocks or intrusive rocks. It has been noted that in the Imperial Valley, California, for instance, there is a correlation between such anomalies and high heat flows. Of course, since anomalies can result from other factors, gravimetric evidence must be supplemented by data obtained through the usage of other exploration techniques.

3.2.5(iv) Seismic Methods
Both reflection and refraction methods can be employed and involve explosions. The former utilises energy reflected from subsurface interfaces between rocks possessing different physical properties. The latter depends upon seismic waves refracted horizontally *along* an interface and thereafter back to the surface. The objective of the two approaches is the same, namely to interpret underground structure and both the configuration and depth of basement rocks. Reflective seismic prospecting has been utilised in order to determine optimal drilling sites for production wells and also to detect subterranean faults and fracture zones. M. Hayakawa has used common-depth-point techniques and analysis of attenuation in a reflection seismic survey at the Matsukawa geothermal area in Japan.[40] Seismic noise is also significant because it is influenced by surface thermal activity; this is discussed in detail in Chapter 11.

3.2.5(v) Microwave Radiometry

This is a valuable technique depending upon the fact that all bodies above absolute zero in temperature emit energy (electromagnetic radiation) and it has an advantage over infrared imagery sensing in that it is capable of penetrating a few metres. Attempts to use the method have been made in Italy, but refinement is necessary.

3.2.5(vi) Satellites

Analysis of data from the Earth Resources Technology Satellite (ERTS-1) shows clearly that with such remote sensing, it is perfectly possible to search for geothermal areas. Imagery from the Skylab Earth Resources Experiment Package (EREP) make further advances possible because of its improved instrument capabilities. Additionally, the Heat Capacity Mapping Mission (HCMM) is a low-cost satellite which is potentially capable of identifying hotter geothermal regions. Actually, data from this latter is to be used in conjunction with the US Geological Survey in order to develop more advanced remote sensing techniques for geothermal exploration. These will involve thermal modelling. No doubt, the Jet Propulsion Laboratory will be involved as it is presently on the continuing research programme on the microwave detection of geothermal anomalies. This Laboratory is also performing research on the enhancement of imagery by digital computers, a technology which has successfully detected mineralised areas at Goldfield in Nevada. Satellite HCMM in polar orbit will be equipped with visible and thermal infrared and heat-capacity sensor also.[41]

Landsat data are also valuable since 2 and C are equipped with thermal infrared scanners. This was particularly interesting to the writer as regards geothermal investigations in Iran where a receiving station at Shahdasht near Karaj is now completed.

3.2.6. Drilling

Under completion of the foregoing exploratory techniques and selection of suitable sites, exploratory drilling is commenced. From this, various important data are assembled. These relate to rate of drilling progress, loss of circulation fluid, temperature and pressure, petrology, stratigraphy, aquifers, the chemistry of the thermal fluid, alteration phenomena, enthalpy and mass flow and permeability. When these results are collated, the volume of the hot reservoir may perhaps be estimated and the drilling properties of the formations listed as well as those of the cap rock. Thereafter, it is possible to estimate also the potential production capacity.

The actual drilling technique is similar to that routinely employed in gas and oil drilling, but certain special problems arise. High temperatures found in geothermal areas may affect the circulatory system and can be bothersome. Such areas often occur in very rugged terrains as is the case at Mount Damavand in Iran and here it is often difficult to bring the equipment into the locality of the drillhole, surveying of which may have been done by helicopter. Naturally, a flat site for this is optimal because it enables the drilling rig to be supported adequately. Often, some sort of road must exist or be constructed in order to facilitate access to the drilling site and enable personnel and tools to get there. Thought must also be given to the availability of utility services, namely power, water, etc., for the actual field operation itself.

The drilling rigs employed need only be of a medium-depth capability because geothermal resources usually lie nearer the surface than is normal for gas or petroleum, i.e. they need to be able to penetrate down only to around 4000 ft or so. The circulation fluid used is normally mud—bentonite in the activities at Damavand designed to investigate the thermal gradient. The actual hole size utilised depends upon the expected steam and/or hot-water volumes. For instance, if the anticipated quantity is 10–25 t/h, then a hole up to 17 in with $13\frac{3}{8}$-in surface casing, fully cemented, may be employed. Casing is inserted to permit the passage of large amounts of steam to the surface and prevents cave-in. At Damavand, the latter occurred in a number of holes and casing had to be inserted at a later stage. Such casing must not only be corrosion-resistant, but also capable of withstanding the effects of friction, rolling, vibration, etc. The latter may be minimised by ensuring as close a fit for the casing as possible. Inside wear results from the steam transfer at high velocity, but outside, iron sulphide in formations may cause deleterious reactions to occur. Stresses induced by high temperatures may also arise. Optimally, the casing ought to be covered by cement. This will also militate against wear. Obviously, in order to combat the effects of high temperature, high-temperature cement must be used. Requirements for the drilling mud are those normally necessary such as the ability to break up rock drilled out by the drilling bit and remove cuttings from the bottom of the well up to the surface, cool and lubricate the bit, control fluid pressure in the formation by means of mud-water column pressure, etc.

The circulation fluid used at Imperial Valley in California was a lignite mud which counteracts high-temperature effects by inhibiting the flocculation of clay, hence acting as a thinner. Here, as also at Damavand, loss of circulatory fluids occurred.[42]

Duration of drilling is variable, but to reach, say, 2000 m, 2 or 3 months may well be necessary and there is usually a subsequent 10-day testing period after that.

The wellhead equipment installed thereafter has to conform to certain specifications tailored to geothermal needs such as

1. ability to handle large flow rates;
2. ability to transmit mud and steam at high temperature;
3. ability to handle relatively low well operating and closing pressures;
4. ability to resist corrosion, both chemical and mechanical, of the mixture of steam, hot water, etc.;
5. ability to be interconnected with large-diameter delivery pipelines.

Safety is of the first importance. For instance, Paul N. Cheremisinoff and Angelo C. Morresi noted that at The Geysers, a blowout blew the top off a hill and, to 1976, remained uncapped.[43] Therefore, a blowout preventer must be installed and this is fitted on to the wellhead and acts as a safety valve by closure on the drill pipe or the casing or directly on the central bore.

Very hard rocks may be encountered during drilling and these may consume a lot of bits and greatly add to costs. More bits may be lost due to the high state of fracturation of most geothermal resource formations. This causes jerking and jumping of the operation unless the bit weight and velocity are very well coordinated. Faults may divert drilling and preliminary geophysical survey is essential to locate them before operations begin. Sometimes, a loss of circulation fluid causes the drilling column to stick where this is fast. Also, high-temperature drilling in cold weather can cause expansion of the pipe joints accompanied by separation of the tool joint from the drilling column. The drill collar may break off as a result of shocks and turbulent drilling. Many of these hazards can be minimised, however, by utilising special techniques of drilling and also by ensuring optimal lubrication. Air drilling may also be utilised and so may turbo-drilling.[44] Of course, it almost goes without saying that there should be continuous inspection during and after the drilling process and the information given in Table 3.2 ought to be collected.[44]

The newly drilled well may be flushed. It may also prove negative and in this event, re-examination of all the geological, geochemical and geophysical data as well as the drilling logs becomes necessary to verify the selection of the well site and determine whether *deeper* drilling or casing perforation might facilitate production.

Drilling in dry geothermal prospects involves the same procedures as

TABLE 3.2

INFORMATION TO BE OBTAINED FROM A WELL

Survey in well	Relevant instrument	Survey purpose
Temperature	Thermometer	Changing temperatures in well; distribution of geothermal heat
Flow	Spinner	Measurement of flow of steam
Pressure	Amerada (Bourdon tube)	Measurement of pressure in well
Inclination	Totoko and Murata type	Measurement of direction and inclination of well
Electric logging	Current and potential electrodes	Comparison of formations
Bottom hole sampling	Bottom hole sampler	Collection of water under pressure in well
Cement bond logging	Improvement of sonic log	Satisfactoriness of cementing

those outlined above, but has the advantage that, initially at least, problems of corrosion due to steam do not arise. The artificial circulation of hot fluids, however, will soon promote it. Also, the same problems of high temperatures and pressures are encountered as in wet drilling.

Borehole information is absolutely essential in order

1. to estimate the ability of the geothermal prospect to produce enough energy over a long enough time to be economically attractive;
2. to distinguish between different models of the geothermal system and thus facilitate accurate prediction of production characteristics under various conditions of exploitation;
3. to calibrate and refine geochemical and geophysical methods for the recognition and delimitation of geothermal systems.

REFERENCES

1. Puxeddu, M., Squarci, P., Rau, A., Tongiorgi, M. and Burgassi, P. D. (1977). Stratigraphic and tectonic study of Larderello–Travale basement rocks and its geothermal implications. Geothermics, 6, 83–93.
2. Combs, Jim and Muffler, L. J. P. (1973). Exploration for geothermal resources. In: Geothermal Energy, ed. Paul Kruger and Carel Otte. Stanford University Press, Stanford, Ca., pp. 95–128.

3. Muffler, L. J. P. and White, D. E. (1972). *Geothermal Energy. The Science Teacher*, **39**(3), 40–3.
4. Boldizsar, T. (1970). Geothermal energy production from porous sediments in Hungary. *Geothermics, Sp. Issue 2*, **2**(1), 99–109.
5. Jones, P. H. (1970). Geothermal resources of the northern Gulf of Mexico Basin. *Geothermics, Sp. Issue 2*, **2**(1), 14–26.
6. Makarenko, F. A., Mavritsky, B. F., Lokchine, B. A. and Kononov, V. I. (1970). Geothermal resources of the USSR and prospects for their practical use. *Geothermics, Sp. Issue 2*, **2**(2), 1086–91.
7. White, D. E. (1968). Hydrology, activity and heat flow of the Steamboat Springs thermal system, Washoe County, Nevada. *US Geol. Surv. Prof. Papers*, **458-C**, 109 pp.
8. Grindley, G. W. (1970). Subsurface structures and relation to steam production in the Broadlands geothermal field, New Zealand. *Geothermics, Sp. Issue 2*, **2**(1), 248–61.
9. Fournier, R. O. (1977). Chemical geothermometers and mixing models for geothermal systems. *Geothermics*, **5**, 41–50.
10. Tonani, F. (1970). Geochemical methods of exploration for geothermal energy. *Geothermics, Sp. Issue*, **2**(1), 492–515.
11. Mahon, W. A. J. (1970). Chemistry in the exploration and exploitation of hydrothermal systems. *Geothermics, Sp. Issue*, **2**(2), 1310–22.
12. Sigvaldson, G. E. and Cuellar, G. (1970). Geochemistry of the Ahuachapan thermal area, El Salvador, Central America. *Geothermics, Sp. Issue*, **2**(2), 1392–8.
13. Fournier, R. O. and Truesdell, A. H. (1970). Geochemical indicators of subsurface temperature applied to hot spring waters of Yellowstone National Park, Wyoming, USA. *Geothermics, Sp. Issue*, **2**(2), 529–35.
14. Dall'Aglio, M., Da Roit, R., Orlandi, C. and Tonani, F. (1966). Prospezione geochimica del mercurio. Distribuzione del mercurio nelle aluvioni della Toscana. *L'Industria Mineraria*, **17**, 391.
15. Matlick, S. and Busick, P. R. (1975). Exploration for geothermal areas using mercury: a new geochemical technique. *Proc. 2nd UN Symp. Development and Use of Geothermal Resources, San Francisco*, **1**, 785–92.
16. Klusman, R. W., Cowling, S., Culvey, B., Roberts, C. and Schwab, A. Paul (1977). Preliminary evaluation of secondary controls on mercury in soils of geothermal districts. *Geothermics*, **6**, 1–8.
17. Brondi, M., Dall'Aglio, M. and Vitrani, F. (1973). Lithium as a pathfinder element in the large scale hydrogeochemical exploration for hydrothermal systems. *Geothermics*, **2**, 142–3.
18. Roberts, A. A. (1975). Helium surveys over known geothermal resource areas in the Imperial Valley, California. *US Geol. Surv.*, Open-file report.
19. Fournier, R. O. and Rowe, J. J. (1966). Estimations of underground temperatures from the silica content of water from hot springs and wet-steam wells. *Amer. J. Sci.*, **264**, 685–97.
20. Sakai, H. and Matsubaya, O. (1974). Isotopic geochemistry of the thermal water of Japan and its bearing on the Kuroko ore solutions. *Econ. Geol.*, **69**, 974–91.
21. Marshall, W. L. and Slusher, R. (1968). Aqueous systems at high temperature,

solubility to 200 °C of calcium sulfate and its hydrates in sea water and saline water concentrates and temperature-concentration limits. *J. Chem. Engineering Data*, **13**, 83–93.

22. Hemley, J. J. (1967). Aqueous Na/K ratios in the system $K_2O-Na_2O-Al_2O_3-SiO_2-H_2O$. *Geol. Soc. Amer. Abstr.*, 94–5.
23. Fournier, R. O. and Truesdell, A. H. (1973). An empirical Na–K–Ca geothermometer for natural waters. *Geochim. Cosmochim. Acta*, **37**, 1255–75.
24. Paĉes, T. (1975). A systematic deviation from the Na–K–Ca geothermometer below 75 °C and above 10^{-4} atm P_{CO_2}. *Geochim. Cosmochim. Acta*, **39**, 541–4.
25. Fournier, R. O. and Truesdell, A. H. (1974). Geochemical indicators of subsurface temperatures—2. Estimation of temperature and fraction of hot water mixed with cold water. *US Geol. Surv. J. Res.*, **2**, 263–70.
26. Truesdell, A. H. and Fournier, R. O. (1977). Procedure for estimating the temperature of a hot water component in a mixed water using a plot of dissolved silica vs enthalpy. *US Geol. Surv. J. Res.*, **5**, 49–52.
27. Arnason, B. (1977). Hydrothermal systems in Iceland traced by deuterium. *Geothermics*, **6**, 125–51.
28. Sakai, H. and Matsubaya, O. (1977). Stable isotopic studies of Japanese geothermal systems. *Geothermics*, **6**, 99–124.
29. McKenzie, W. F. and Truesdell, A. H. (1977). Geothermal reservoir temperatures estimated from the oxygen isotope compositions of dissolved sulfate and water from hot springs and shallow drillholes. *Geothermics*, **6**, 51–61.
30. Sakai, H. (1977). Sulfate–water isotope thermometry applied to geothermal systems. *Geothermics*, **6**, 67–74.
31. Bowen, R. and Williams, P. W. (1972). Tritium analyses of ground water from the Gort Lowland of western Ireland. *Experientia*, **28**(5), 497–8.
32. Bowen, R. and Williams, P. W. (1973). Geohydrologic study of the Gort Lowland and adjacent areas of western Ireland using environmental isotopes. *Bull. Water Resources Res.*, **9**(3), 753–8.
33. Kruger, P., Stoker, A. and Umana, A. (1977). Radon in geothermal engineering. *Geothermics*, **6**, 13–19.
34. Mazor, E. (1977). Geothermal tracing with atmospheric and radiogenic noble gases. *Geothermics*, **6**, 21–36.
35. D'Amore, F. and Nuti, S. (1977). Notes on the chemistry of geothermal gases. *Geothermics*, **6**, 39–45.
36. White, Donald E. (1973). Characteristics of geothermal resources. In: *Geothermal Energy*, ed. Paul Kruger and Carel Otte. Stanford University Press, Stanford, Ca., pp. 69–94.
37. Hodder, D. T. (1970). Application of remote sensing to geothermal prospecting. *Geothermics*, *Sp. Issue*, **2**(2), 368–80.
38. Krynine, Dimitri P. and Judd, William R. (1957). *Principles of Engineering Geology and Geotechnics*. McGraw-Hill Book Company, New York, Toronto, London, 730 pp.
39. Dey, Abhijit and Morrison, H. F. (1977). An analysis of the bipole–dipole method of resistivity surveying. *Geothermics*, **6**, 47–81.
40. Hayakawa, M. (1970). The study of underground structure and geophysical

state in geothermal areas by seismic exploration. *Geothermics, Sp. Issue*, **2**(2), 347–57.
41. World Meteorological Organization (1976). *Informal Planning Meeting on the Satellite Applications in Hydrology.* WMO, Geneva, 25–27 October 1976, Final Report, 35 pp.
42. Cromling, J. (1973). Geothermal drilling in California. *J. Petroleum Technology*, **25**, 1033–8.
43. Cheremisinoff, Paul N. and Morresi, Angelo C. (1976). *Geothermal Energy Technology Assessment.* Technomic Publishing Co. Inc., Westport, Conn., 164 pp.
44. Matsuo, K. (1973). Drilling for geothermal steam and hot water. In: *Geothermal Energy*, ed. H. C. Armstead. UNESCO, Paris, pp. 73–83.

CHAPTER 4

Observations on Hydrothermal Convection Systems

The physico-chemical properties of all potentially exploitable concentrations of geothermal heat near the planetary surface are interesting,
but the main usable resources which are recoverable using existing
technical and economic means are to be found among the hydrothermal
(i.e. vapour-dominated or hot-water-dominated) convection systems. In
these, heat transfer is primarily through circulating fluids, not by
conduction, and the convective process results from the heating of fluids in
a gravity field, a process causing thermal expansion. Those of low density
tend to ascend and be replaced by cooler, higher-density fluids so that
convection clearly increases temperatures in the upper portion of a system
while lowering them basally. By its very nature, it disturbs conductive
gradients and consequently, no one temperature gradient or heat flow is
able to characterise a convective system. In fact, surficial gradients as high
as 3 °C per metre of depth have been recorded and it is obvious that they
must greatly diminish with depth.[1] Actually, this has been shown to be the
case where drilling has been effected.

Hydrothermal convective systems have been discussed in Chapter 2 and
the present discussion is an amplification intended to indicate their
characteristics in greater detail.

4.1. VAPOUR-DOMINATED SYSTEMS

Some of the most important geothermal systems in the world, for instance
The Geysers of California and the Larderello fields in Italy, produce dry or
superheated steam without any associated liquid and are sometimes
referred to as 'dry-steam' systems. D. E. White, L. J. P. Muffler and A. H.
Truesdell, however, believed that both liquid water and vapour must

coexist in the reservoir with vapour as a continuous and pressure-controlling phase.[2] If this is indeed the case, then the term 'vapour-dominated' seems highly appropriate. The actual nature of the initial fluid at Larderello has been the subject of much speculation. Suggestions include

1. saturation with liquid and superheated steam in a reservoir which, while producing, is replenished by boiling from a deep-water body,[3] this perhaps being a brine;[4]
2. a reservoir largely filled with vapour, but possessing local disturbance waters;[5]
3. a hot-water system with very low recharge rate and consequent loss of water by boiling.[2]

Two subtypes of the vapour-dominated system are distinguishable, namely the Larderello and the Monte Amiata.

4.1.1. Larderello Subtype
The Larderello subtype (v. Fig. 4.1) possesses the following characteristics, physically, chemically and geologically exemplified at Larderello, The Geysers and Matsukawa.

1. Initial temperatures near 240 °C in reservoirs at or lower than 350-m depth. Here, it is important to note that isotopic geothermometers based on $^{13}C/^{12}C$ fractionation between CO_2 and CH_4 and the $^{18}O/^{16}O$ fractionation between CO_2 and H_2O vapour have been applied at Larderello by C. Panichi and his associates.[6]
   The CO_2–CH_4 thermometer gave temperatures 50–200 °C *higher* than those measured at the wellhead. However, the distribution of the isotopic temperatures within the field follows the same or at least roughly similar patterns to those derived from wellhead temperature measurements and they probably reflect the temperatures of formation of CO_2 and CH_4. The CO_2–H_2O thermometer gives the temperatures of the tapped reservoir which may be higher than those indicated above. No doubt, the difference between isotopic temperatures and those which are measured at the wellhead result from the cooling undergone by the geothermal fluids *en route* to the surface.
2. Initial temperatures and pressures are believed to be quite uniform and to be influenced by the maximal enthalpy of saturated steam which is 669·7 cal/g at 236 °C and 31·8 kg/cm² according to R.

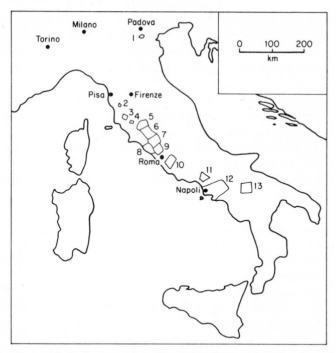

FIG. 4.1.   Some of the important geothermal regions undergoing investigation in Italy. 1 = Monte Berici; 2 = Monte Catini; 3 = Larderello Travale; 4 = Roccastrada; 5 = Monte Amiata; 6 = Monte Volsini; 7 = Monte Cimini; 8 = Tolfa; 9 = Monte Sabatini; 10 = Colli Albani; 11 = Roccamontina; 12 = Campania Ovest; 13 = Monte Vultura.

James[3] and others.[2,5] As the gas content of the vapour increases, however, there are marked changes.

3.   In vapour-dominated reservoirs of this subtype, pressures are below hydrostatic.

4.   Where surface activity is intense, manifestations such as fumaroles, mud pots and volcanoes and acid-leached ground occur, and springs in the region are usually acidic because of H_2SO_4 produced by oxidation of escaping gas and have pH values as low as 2 or 3 (pH measurements are very useful in geothermal areas). Sometimes, neutralisation by ammonia may occur. While sulphates may be high, chlorides are low. Where such intense surficial manifestations do not occur, the spring waters are slightly acid only, if not actually alkaline.

5. It has been noted that discharge rates of fluids from vapour-dominated systems are consistently low with a range of from some tens to several hundred litres per minute.
6. The heat is probably mostly in solid phases, perhaps as much as 90%.[1]
7. While the production wells produce dry to superheated steam, liquid water has been suggested as occurring in non-commercial wells peripheral to reservoirs as well as in fluid initially produced from some wells which alter from wet steam to dry steam.

As production proceeds at Larderello, there has been a steady increase in wellhead temperatures, most wells thereafter commencing to decline. Enthalpy has risen to as much as 710 cal/g together with increasing superheat.[5] This may be the result of either

1. an increasing dependence upon supercritical water (over 374 °C) from deep magma,[5] or
2. an increasing dependence upon boiling from a deep-seated and declining saline water body which becomes more saline as water is vaporised.[1]

Systems of the above-mentioned subtype, e.g. The Geysers, Larderello and Matsukawa, are characterised by initial temperatures of at least 240 °C, shut-in pressures near 35 kg/cm^2 and gas contents (excluding steam) of less than 5%. D. E. White concluded that discharge areas are vital for such systems because they allow for the net loss of a great deal of initial pore water and therefore permit the domination of the vapour phase as well as the flushing out of gases other than steam.[1]

4.1.2. Monte Amiata Subtype
The Monte Amiata subtype resembles hot natural-gas fields in some respects. For comparable initial pressures (20–40 kg/cm^2), temperatures are much lower (150 °C) and initial gas contents much higher (over 90%) than in the Larderello field.[1] Steam is a rather small initial constituent and, following production and decompression, the initial vapour of high content is flushed out of the reservoir and is replaced by relatively low-pressure steam of lower gas content resulting from the boiling of water at moderate temperatures. R. Cataldi noted another feature of the subtype, namely a trend in fluids produced from the Bagnore field of the Monte Amiata district to go from dry vapour to vapour and liquid water.[7] This may result

from water-flooding or possibly a rising deep water table responding to the surrounding hydrostatic environment as the reservoir pressures go down with production. The greatest obstacle to fluid flow in this subtype most likely takes place in the discharge part of the system as a result of low-permeability cap rocks.

4.2. HOT-WATER SYSTEMS

In hot-water systems, liquid water is the continuous, pressure-controlling fluid phase.[2] Of course, some vapour may also be present, sometimes as bubbles in the shallow, low-pressure regions. The normal constituents of water analysis are present and include silica, alkali metals such as sodium and potassium, calcium, magnesium, Cl, SO_4, HCO_3 and CO_2. Water is the transmitting agent for heat and the rapidity of the transmittal depends largely upon permeability. The heat is acquired at depth by recharge and other waters through conduction from hot rocks deriving their heat from magma.

D. E. White has listed several subtypes which may be appended:[1]

1. Moderate temperature systems (50–125 °C) with waters chemically similar to surface and shallow ground waters of the area.
2. Systems in deep sedimentary basins usually saline water-bearing and of moderate temperature with waters which are certainly partially non-meteoric in origin. Some instances have been located in the California Coast Ranges.
3. Hot-water systems with very highly saline brines such as the Salton Sea geothermal system and the Red Sea brine pools about which H. Craig and D. A. Ross have written.[8,9] As the present writer and A. Gunatilaka have pointed out, such areas are also incipient ore deposits, at least as regards copper.[10] The brines are about 26 % salinity in both cases, but there are marked physico-chemical differences.
4. Some systems have natural cap rocks which inhibit discharge and incidentally insulate the reservoirs, thus conserving their heat. Both the Salton Sea and the Cerro Prieto system in Mexico have low-permeability cap rocks.
5. Self-sealing, i.e. creation of its own insulating cap rock, may occur in some high-temperature, hot-water systems. This takes place

when hydrothermal minerals are deposited in pore spaces near the surface and the phenomenon has been recorded in the Yellowstone National Park, Wyoming, USA, and also at Wairakei, New Zealand.

It will be useful now to add the general properties of hot-water systems.

It could be thought that hot springs always indicate a geothermal reserve, but this is not always the case. Analyses show that the highest temperature springs are, where discharge rates are also high, also the highest in the SiO_2, Cl, B, Na, K, Li, Rb, Cs and As contents as compared with environmental ground water. In the last chapter, the use of some of these as chemical geothermometers was indicated. It is believed that there is a base temperature which is characteristic of the deeper parts of many hot-water convection systems and, in systems with fairly low salinity, this may rise to as much as 300 °C. Of course, the Salton Sea brines are hotter than this, attaining 360 °C, and the Cerro Prieto system in Baja California (possessing a salinity two-thirds that of sea water) can be as hot as 388 °C, according to various workers. Temperatures of this order are above the critical temperature of pure water and can occur in brines as a result of the fact that their physical properties are different from those of pure water.[11]

It appears that silica is the most significant self-sealant of hot-water systems. In low-temperature periphery areas and self-sealed cap rocks, opal and beta-cristobalite are characteristic, according to S. Honda and L. J. P. Muffler.[12] Clay minerals, calcite and even zeolites may also be important.

Where geysers occur, for instance in the Yellowstone National Park, direct evidence is available of the upflow of subsurface waters with temperatures of 180 °C or more. Recrystallised or amorphous silica is also evidence of this. On the other hand, travertine implies *low* subsurface temperatures, a fact of great interest to the writer because the Damavand region contains many deposits of this $CaCO_3$ form in association with springs.

Where convection systems are low temperature, of course, there is not much chance of self-sealing taking place because there is a paucity of silica in the waters. High-temperature systems having maximal temperatures in excess of 180 °C tend to decrease in permeability with time, especially in the upper parts, because of self-sealing.

Wells in permeable reservoirs usually produce anything from 70 % to 90 % of total mass flow as water, the proportion of steam produced when pressure is reduced being related to initial fluid temperature and also to final separating pressure. Wells occurring in low permeable areas may at first emit water and steam and later this changes to wet steam and ultimately becomes dry steam. Increasing steam content is not a blessing because it implies evaporation of all local water, temporarily producing steam and also precipitating dissolved matter and thus decreasing permeability.

In the majority of hot-water fields, the fluid which enters a producing well is liquid water and this remains so as it ascends until such time as steam bubbles commence to form. As the upward flow goes on, more water flash boils to steam as the pressure and mixture temperature go down. Steam is buoyant and displaces residual water upwards, thus accelerating the emission of the mixture and also ejecting the water above ground as in a natural geyser. Where first boiling takes place in a well depends upon the initial temperature of the water and also on formation-fluid pressure and separator pressure.

Chloride contents of water which has been at a temperature exceeding 150 °C or so are almost always higher than 150 ppm. Chloride is an important parameter in distinguishing hot-water systems from vapour-dominated ones. The majority of metal chlorides are highly soluble in water (liquid) and also chloride in most rocks is very easily leached out by hot waters. However, the common metal chlorides have quite negligible volatility at temperatures even up to 400 °C and cannot dissolve in low-pressure steam. Consequently, a chloride-bearing water (over 50 ppm) is a definitive index of a hot-water system. There are a number of hot-water systems, however, which produce surficial acid–sulphate springs with *low* chloride contents and these are sustained by steam boiling from some underlying chloride-bearing water body. Otherwise, they resemble chemically springs associated with vapour-dominated systems.

4.3.   EXPLOITATION

One of the greatest problems in using geothermal energy fields on a large scale is the fact that the most suitable type, namely the Larderello subtype of vapour-dominated system, is very rare. In fact, probably not more than 5 % of all geothermal systems with temperatures above 200 °C belong to it. Nevertheless, of all existing power-generating capacity, almost three-quarters originated in this subtype in 1973.[1]

It is believed that a discharge area is necessary for the subtype to be recognised by its surficial manifestations. The Monte Amiata subtype, perhaps commoner, is more difficult to identify precisely because it lacks these surface manifestations on a conspicuous scale. Actually, other factors militate against its exploitation and these relate to its production characteristics.

Corrosion is apparent as one of the problems which may arise in exploitation, silica and calcium carbonate sometimes forming scale in wells and surface pipes after flash eruption and cooling; where these phenomena occur in reservoirs near wells, both permeability and production rates may be greatly reduced. Some hot waters are corrosive also because they are highly saline or contain H_2SO_4 or have high CO_2 or oxygen contents. In a few cases, HCl may be present also. Recently, corrosion and mineral precipitation have been, at least to some extent, overcome by the use of heat exchangers and special materials, and also by maintaining back pressure in order to prevent flashing in the pipes.[13] Waste disposal is now being handled in new installations by returning fluids to the reservoir. With H_2S which is noxious, toxic and often present in non-condensable gases, chemical adsorption methods are being tried. These would eliminate the pollution effects of dispersal in the atmosphere. Reinjection techniques are frequently suggested in order to deal with those hot-water effluents which involve an environmental hazard. However, the reliability of this must be carefully tested because sometimes hot-water effluents may not be compatible with reservoir fluids even though they were initially the same. Reinjection was used in the Long Valley system in California and also, for a year, in the Salton Sea system. The only fairly long-term experiment until recently was at The Geysers, however, where, over 3 years to 1972, cool condensate was reinjected successfully into an underpressured vapour-dominated reservoir. Vapour-dominated systems such as The Geysers and Larderello have at present electrical power capacities of about 900 and 405 MW respectively, not a very big increase on the 1972 figures presented in Table 2.8. Development plans for The Geysers, however, are expected to double the capacity there within a few years and the ultimate capacity may rise to anything between 1200 MW and 4800 MW! The Larderello field is now practically at capacity, but new Italian discoveries elsewhere will continue to enlarge the total capacity there.

Steam at about 7-bars pressure is admitted to turbines at The Geysers plant and exhaust steam is condensed by direct contact with the cooling water. Cooling towers are employed and evaporation reduces to some 20% the quantity added by the condensed steam. This overflow is then returned to the geothermal reservoir by means of a disposal well. By gravity alone, the return is effected and this is a consequence of the fact that reservoir pressures are well under hydrostatic. Non-condensable gases are exhausted from the condenser into the atmosphere. Recently, it has been decided to take out hydrogen sulphide, however.

The main installations using fluids from high-enthalpy, liquid-dominated systems are those at Wairakei, New Zealand and Cerro Prieto, Mexico, although smaller amounts of energy are produced at Otake, Japan, Pauzhetsk, USSR, and Namafjall, Iceland. In another New Zealand locality, namely Kawerau, 10 MW is produced in order to operate an industrial plant. In all these, steam has to be separated from the steam–water mixture which is derived from the wells. The steam pressures at the turbines vary from one to another, but remain in the area of 3 or 4 bars. At Wairakei, surface condensers are employed because there is plenty of river water available for cooling and both the flashed fluid from the separators and the condensate are transmitted to the river without any apparent ill effects upon the environment. At Cerro Prieto, direct condensation and cooling towers are utilised, the flashed fluid and overflow being conveyed to the Gulf of California. Of course, monitoring has not been effected over a long enough period of time to be sure whether or not such outputs can be deleterious to the environment or not and every care is therefore advisable.

Steam may be used not only for the generation of electrical power, but also for other industrial applications.

Low-enthalpy geothermal fluids may have a great future potential and it is to be hoped that this is indeed the case because they are more abundant than high-enthalpy ones. It is quite feasible to transmit geothermal energy from such resources through correctly insulated lines over quite long distances. It has been stated that in such a line, a fluid at 150 °C could be conveyed 50 km with a maximal temperature loss of 25 °C.[13] Heat transmitted in this manner could be used for many purposes, space heating being one. In fact, this is being done in Iceland, half the population there receiving heat from such sources. The capital, Reykjavik, gets almost *all* its heating requirements from hot-water reservoirs at temperatures of from 90 °C to 150 °C as far away as 18 km. In Iceland, there are other applications too, for instance greenhouse heating and the drying of diatomite. Low-enthalpy waters are similarly utilised in Hungary and the USSR as well as in the USA where both Boise, Idaho, and Klamath Falls, Oregon, employ them for residential, school and business space-heating requirements. Mostly, the wells at Klamath Falls use a heat exchanger in the well in order to comply with city ordinances that hot waters may not be discharged into the sewer system. Passing into the higher temperature range for these low-enthalpy fluids, i.e. 160–200 °C, the only instance at present of use in electric power generation is in Kamchatka, USSR, where a 680-kW freon unit is in operation.

One of the uses to which both low-enthalpy and high-enthalpy geothermal fluids may be put is in providing fresh water. Both the multiple effect and the multistage flash distillation processes are efficient in utilisation of heat and in either, the geothermal fluid may constitute the feed or its heat may be exchanged with another impure water as the feed. Which method is used depends upon problems of corrosion and sealing. Investigations are currently being made in order to ascertain whether, in the case of the geothermal fluids of the Imperial Valley in California, desalination is feasible or not.

In the case of Italy, geothermal energy was utilised for centuries and the first attempts to employ it for purposes other than heating bathing pools were actually made in the Larderello area as early as the late eighteenth century. Steam from fumaroles was used in boric acid extraction from hot pools. The long history of the Larderello fields has been characterised by much innovation. One is the installation of rather small (1·5–5 MW) back-pressure turbines exhausting directly to the atmosphere and these are utilised on individual wells early in the development of a new field. There are a number of advantages, for instance they can handle steam containing large amounts of non-condensable gases such as CO_2, thus releasing them and improving the steam ratio until it can be used in conventional condensing turbines. A second advantage is that the temperature–pressure–volume relationships of the reservoir can be determined by production testing and thereafter predictions can be made regarding the life of the reservoir prior to funds being committed for further development. Another important innovation has been the development of the technique of using air as a circulation fluid in drilling geothermal wells. This way, it is possible to drill into low-pressure steam zones without their becoming sealed off with a cake of mud. In fact, it has now become standard practice to employ air in all vapour-dominated systems and the method has been found to increase greatly the flow per well because more zones are able to produce steam.

Initiation of attempts to develop the geothermal resources of the North Island of New Zealand were made in the early 1930s, but serious studies only really began after the second world war, studies aimed at construction of a geothermal power station. By 1953, drilling in the area of Wairakei had demonstrated that sufficient steam was available for this purpose and the first station was actually completed in 1960, a second being added in 1963.

4.4. ECONOMICS

This book will not examine the economics in any great detail, but it is certainly necessary to say something about them.

Initial outlay as regards exploration and later expenditures on the development of geothermal fields and installing heating systems have discouraged the use of derived energy in space and process heating. This is not the case in places where the geothermal resources are readily available and known for a long time such as Iceland, but generally speaking, this is not the case. In order to produce electrical power from geothermal sources, however, it is a fact that the costs are either similar to or lower than those involved if other methods (except hydroelectric) are employed. Another advantage of geothermal energy is that it is renewable, unlike our fossil fuels. The costs of these latter have shown a steady increase throughout the century and, in recent years, have indeed skyrocketed. Additionally, there is ever-increasing concern about the effects on the environment of continually rising consumption. Consequently, it is becoming apparent that geothermal energy presents an attractive alternative because it is an assured energy supply and has a lower operating cost. It has been shown that, where geothermal systems have been developed for heating, their costs range between 10 and 25 % of the cost of fossil-fuel systems.[13] In Iceland, Gunnar Bödvarsson and Johannes Zoëga quoted figures ranging from US7·5c. to US12c. per million British thermal units for district heating in the area of Reykjavik; imported oil heating cost $US1·02 per million British thermal units for comparison.[14] These remarks were written and published in 1964, but the relation is probably more favourable than would be the case today.

A report made by W. D. Purvine is more relevant perhaps, because it was published in 1974.[15] Costs for geothermal heating at the Oregon Institute of Technology located in Klamath Falls were discussed. This is a new campus with eight detached buildings covering a total area of about 400 000 ft^2 or more. The cost of geothermal heating there was, averaged over the period since opening, about $US10 000–40 000 annually. These may be compared with costs at an older campus heated by oil-fired hot water which averaged $US94 000 annually before the move to the new campus in 1964 was effected.

A number of people have reported costs from producing electrical power from geothermal sources and all show that both the capital and the operating costs are lower than those for other base-load thermal plants. Nowadays, the cost of constructing a geothermal plant is substantially lower than that of constructing a comparable fossil-fuel plant and as low as only one-third or one-quarter of the cost of building a nuclear plant. Operation costs are also lower proportionately and it is rather a simple matter to convert to or install initially automatic and unattended operation. The Geysers, in fact, normally operates unattended for two-thirds of the day and maintenance personnel are only on duty during the day itself. All these factors mean that geothermal plants can be constructed to give smaller outputs than conventional or nuclear plants where, in order to achieve lower costs, plants producing hundreds of megawatts must be built. The result of this is that utilities sometimes overbuild and then have to wait until demand catches up with capacity, not a very efficient business method.

In recent years especially, following dimouts and blackouts and their accompanying crime and baby booms, realisation has set in of the utter dependence of modern society upon a constant source of electricity. Political factors in many parts of the world have influenced the situation, usually adversely, through withholding of fossil fuels or meteoric price increases in attempts to blackmail western consumers. Great economic pressures have resulted. Where they occur, geothermal systems can help to overcome these difficulties. Their heat is constant and alters only over what is, by human standards, very long time intervals. The operation of such resources is independent of outside support because the geothermal fluids flow to the surface by using a part of their own energy and thereafter pass directly into the power plant. No rail connections are required, no mines exist, no processing plant is necessary. The situation is one which almost obviates the possibility of strikes and, an almost equally catastrophic phenomenon, political decisions! All these facts make geothermal systems very reliable and this reliability is such an important aspect that it rivals the advantage of low costs.

4.5. BREAKTHROUGHS

It is to be expected that advances in utilisation technology will take place and enable a great expansion in the development of geothermal systems to occur. One of the most significant advances is likely to be an improved heat-exchange technology which would allow the heat to be used from fluids at a temperature of 100 °C or below.[16] This is because the amount of heat available in recoverable natural fluids in the lower temperature range is far more than that available from fluids at higher temperatures, say 180 °C or above. Another advance of great importance will be that which enables cheap artificial fracturing of hot, dry rocks to be effected efficiently and thus facilitate both the introduction of fluids as a result of adequate permeability and a consequent recovery of stored energy. Of course, this is a big subject and it is discussed in more detail later.

Other advances which would be extremely useful include the devising of better drilling technology in order to facilitate the drilling of holes cheaply to great depths, the linking of geothermal energy development in some way to desalination which would again result in a lowering of costs, and new technology development which could widen the applications of such energy into fields such as horticulture and product processing. If all these breakthroughs are achieved, there would be tremendous advantages. For instance, geothermal energy could be recovered from large-scale gradient-dominated rock masses such as the deep sedimentary basins as well as from vast regions of hot and dry crystalline rocks.

However, D. E. White has opined that the total planetary geothermal power production is not likely to exceed 30 000 MW at present prices and with existing technology.[1] He stated that proved, probable and even possible reserves in the USA which are recoverable comprise about 600 MW-centuries. If one-third more money was spent, however, this figure would rise to perhaps as much as 4000 MW-centuries, even with the use only of existing technology.

More cash and more research are clearly highly desirable. At the present time, there are something like 10 really productive geothermal regions in the world and they produce a combined output under 1000 MW, a quantity totally inadequate for even the needs of the USA alone. This emphasises the nature and dimensions of the problem.

REFERENCES

1. White, Donald E. (1973). Characteristics of geothermal resources. In: *Geothermal Energy*, ed. Paul Kruger and Carel Otte. Stanford University Press, Stanford, Ca., pp. 69–94.
2. White, D. E., Muffler, L. J. P. and Truesdell, A. H. (1971). Vapor-dominated hydrothermal systems compared with hot-water systems. *Econ. Geol.*, **66**, 75–97.
3. James, R. (1968). Wairakei and Larderello. Geothermal power systems compared. *N.Z. J. Sci. Tech.*, **11**, 706–19.
4. Craig, H. (1966). Superheated steam and mineral-water interactions in geothermal areas. *Trans. Amer. Geophys. Union*, **47**, 204–5.
5. Sestini, G. (1972). Superheating of geothermal steam. *Geothermics, Sp. Issue*, **2**(2), 622–48.

6. Panichi, C., Ferrara, G. C. and Gonfiantini, R. (1977). Isotope geothermometry in the Larderello geothermal field. *Geothermics*, **6**, 81–8.
7. Cataldi, R. (1967). Remarks on the geothermal research in the region of Monte Amiata (Tuscany, Italy). *Bull. Volcanol.*, **30**, 243–70.
8. Craig, H. (1966). Isotopic composition and origin of the Red Sea and Salton Sea geothermal brines. *Science*, **154**, 1544–8.
9. Ross, D. A. (1972). Red Sea hot brine area: revisited. *Science*, **175**, 1455–7.
10. Bowen, Robert and Gunatilaka, Ananda (1977). *Copper: Its Geology and Economics*. Applied Science Publishers Ltd, London, 366 pp.
11. Haas, J. L. Jr (1971). The effect of salinity on the maximum thermal gradient of a hydrothermal system at hydrostatic pressure. *Econ. Geol.*, **66**, 940–6.
12. Honda, S. and Muffler, L. J. P. (1970). Hydrothermal alteration in core from research drill hole Y-1, Upper Geyser Basin, Yellowstone National Park. *Amer. Mineralogist*, **55**, 1714–37.
13. Bolton, R. S., Bowen, R. G., Groh, E. A. and Lindal, Baldur (1977). Geothermal energy technology. Section 7 in: *Energy Technology Handbook*, ed. Douglas M. Considine. McGraw-Hill Book Company, New York, pp. 1–57.
14. Bödvarsson, Gunnar and Zoëga, Johannes (1964). Geothermal energy. In: *Proc. UN Conf. New Sources of Energy: Solar Energy, Wind Power and Geothermal Energy*, Rome, August 21–31, 1961, pp. 2–3 (pub. 1964).
15. Purvine, W. D. (1974). Utilization of thermal energy at Oregon Institute of Technology, Klamath Falls, Oregon. In: *Proc. Intern. Conf. Geothermal Energy for Industrial, Agricultural and Commercial–residential Uses, Klamath Falls, Oregon*, October 7–9, 1974.
16. Jonsson, V. K., Taylor, A. J. and Charmicheal, A. D. (1969). Optimisation of geothermal power plant by use of freon vapour cycle. *Timarit VFI*, **54**, 2–17.

CHAPTER 5

Artificial Stimulation of Geothermal Systems

The necessity for artificial stimulation of geothermal systems is apparent. Formations exist which cover huge areas and contain vast volumes of hot but dry rock having abnormally high thermal gradients and some of these are believed to attain temperatures between 300 and 600 °C and at quite shallow depths of under 3 km. In the USA, most areas of geothermal energy have average formation temperatures exceeding 150 °C within a depth of 10 km and these could be made productive by introducing surface water under pressure. As regards steam and hot-water systems, the dry-steam variety produces electrical power more efficiently than any other type of geothermal resource. However, fields like The Geysers and Larderello have demonstrated that, with the passage of time, a decline in steam pressure from producing wells is taking place. Attempts to halt this involve injection of surface waters or the reinjection of condensate from the turbine exhausts, but they will not arrest it. Increased permeability very well might and this can possibly be achieved by using small charge, optimally spaced, high-energy chemical explosives. The efficacy of these will be related to the very varying conditions found in the various geothermal resource locations. For instance, at Larderello, this derives from anhydrite–dolomite and limestone whereas at The Geysers, graywackes are involved and in Wairakei volcanic debris. C. F. Austin, W. H. Austin Jr and G. W. Leonard have listed five fundamental models of geothermal deposits, namely granite-stock heat sources, basaltic-magma heat sources, wet geothermal-gradient heat sources, metamorphic-zone heat sources and dry geothermal-gradient heat sources.[1] An eastern California granite-stock area was selected for an examination of the ways in which explosive stimulation might be employed. Bore-hole enlargement is one such application, i.e. springing a hole for the purpose of making a larger diameter hole at a chosen horizon. This can create a cavity several times larger than the original diameter of the bore-hole. Explosives require only a light service rig and there is little chance of trapping debris or equipment in the hole. Also, the vibration and shock resulting from the explosion can actually be useful in that they can improve production rates. Structural factors are significant. In a granitic or metamorphic host, an effort to enlarge a bore-hole by under-reaming may end in loss of tools and occurrence of a fine-grained dyke of considerable strength might offer serious eccentric loads on any tool which rotates. Explosives would solve these problems and springing is optimally effected utilising brissant explosives. Obviously, the technique has to be used cautiously in thin-bedded and steeply dipping horizons. Blasting will destroy grain-to-grain bonding as well as physical blockages of fractures and pores and may also very well create new fractures. Permeability may be increased by stress waves and also from perforation, i.e. a conical-shaped charge, one which has been used often in place of bulk-explosive charges in order to increase the effective diameter of a well by means of perforations at right angles to the axis of the bore-hole. Conical-shaped charges can perforate casing and cement in order to give access from the reservoir to the production tubing and also they can increase the effective diameter of a well under the right conditions by as much as a factor of 3. Bullets have also been employed in perforating rock in place of the jet from a conical-shaped charge. Explosive stimulation of a geothermal well requires understanding of several factors, among them

1.   the means of delivery;
2.   the type of host rock involved;
3.   the fluid phase in the formation, i.e. steam or gas or liquid.

Detonating an explosive charge in a bore-hole results in two recognisable zones in the host rocks, a close-in zone in which the fabric of the rock is destroyed as a result of crushing and compaction and a distant zone in which the grain-bond failures and fracturation occur. In this latter, stress-wave passage is the primary phenomenon and this has been studied by several workers such as W. Goldsmith, C. F. Austin, H. C. Wang and S. Finnegan.[2] They found, for instance, that basalts subjected to these show slight attenuation and virtually no dispersion, the coefficient of the former and Young's modulus tending to decrease as the number of shocks experienced by a sample at a given initial shock level increases. Strength does not appear to be affected and basalt turns out to be a good example of the capacity of a brittle material to reach damage saturation with a given severity of repeated impacts, implying that attempts to produce fracturation in a basalt reservoir would not be successful beyond a single detonation.[3]

Stress-wave-induced damage has been found to be localised in porous media and tests on scoria have demonstrated that it and comparable materials are excellent energy-absorbing media.

As to the explosives, charges may be constructed for temperatures as high as 400 °C. At greater temperatures, however, the cost becomes prohibitive. Of course, the explosive selected must be compatible with the dimensions of the job and the conditions under which utilisation is intended. Some of the geothermal-well conditions which act as determinants of the selection of the explosive to be used include the side effects of temperature such as melting which is capable of changing the expected explosive geometry. Another factor is the possibility that well gases or fluids may very well interact with the explosive and perhaps sensitise or desensitise the latter. Complete isolation of the explosive from such gases or fluids is obviously desirable. In this connection, it is necessary also to bear in mind the possibility of interaction between the explosive and materials utilised in well-boring such as lubricants. Another phenomenon which must be examined is the possibility that toxicity or contamination may take place after the shot is carried out and this means that charges with high outputs of toxic chemicals such as mercury must be controlled rigorously with especial attention being paid to amounts employed. Disposal of failed or unused charges must also be considered. There are two possible approaches, namely the use of uncased and the use of cased explosives.

1. Uncased explosives: When in liquid form, these establish a close wall contact and may also be injected into rock pores. Prediction of their behaviour, however, is not at all easy. It has been found, for instance, that in the metal-mining industry, use of some types of slurried explosive in drill holes which possess pyrite in their walls causes an apparent instability. Another drawback in the use of uncased explosives is that it is very expensive to obtain them in a form which can survive temperatures up to 400 °C. A heat-resistant explosive available from DuPont may be mentioned and this is TACOT.

Density and solubility of uncased explosives are properties which can cause difficulties to arise because they may cause them to float or to sink within the bore-hole.

2. Cased explosives: These are much more predictable, of course, and therefore attractive in stimulation work. They can be pre-designed and supplied with full operating instructions thus simplifying matters and ensuring that no particular expertise is necessary on site. TNT is readily available, is rather cheap and also fairly stable. TNT melts above 80 °C, but the liquid can be set off using a lead-azide detonator and an HMX booster. As seen earlier, the vast majority of geothermal wells fall into a temperature range of from 50 °C to 200 °C and TNT is applicable and needs only a very light casing in order to prevent the intermixing of the well fluids and the explosive. Of course, some wells tap higher temperature fluids and here TNT may still be utilised in conjunction with an insulated case. However, this may modify the pattern of energy transfer into the walls of the bore-hole and could contribute more debris also.

C. F. Austin and G. W. Leonard referred to the tests on TNT feasibility at such temperatures effected at the Naval Weapons Center in 1972.[3] Teflon casing was employed because this material has a low thermal diffusivity, is compatible with TNT and can resist rapid decomposition at such temperatures. Additionally, its use avoids introduction of metallic debris into the well. Results showed that selection of an appropriate thickness for the Teflon wall guaranteed reasonable handling times and also facilitated the emplacement and detonation of TNT charges in unquenched wells at up to 400 °C. Austin and Leonard suggested that, where longer times are required than can be achieved using encased TNT, in the temperature range 200–300 °C, a main charge of TACOT could be employed. The TACOT would act as a heat sink and insulator for a lead-azide detonator with a NONA booster and in the temperature range 300–500 °C, main charges of NONA could be utilised up to 400 °C. In order to protect the detonator, a vented ablative system is necessary.

5.1. NUCLEAR EXPLOSIVES

Nuclear explosives will become very significant because of the fact that their application ensures high subterranean fracturing efficiency capable of effective stimulation of geothermal resources areas so that they can increase electricity generated to a point at which their output might make a substantial impact upon the energy needs of the relevant countries. In 1957, the United States Atomic Energy Commission (USAEC) established the Plowshare Program and subsequently, this body accumulated extensive theoretical and experimental information regarding underground nuclear tests, hundreds of which have been conducted after the Limited Test Ban Treaty was signed in August 1963. Yields have sometimes been very small,

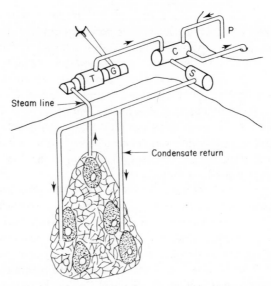

FIG. 5.1. An illustration of the project Plowshare concept, i.e. fracturation of subterranean hot, dry rocks underlying 10 western states in the USA by means of hydrogen bombs followed by piping in of water to be converted to steam below 2 miles depth at a temperature of approximately 180 °C. This steam would flow to the surface through the fractures, would be super-heated and would be piped subsequently to a turbine generator, the condensate being recirculated back to the underground cavity. T = turbine; G = generator; C = condenser; S = surge tank; P = cooling pond.

under 1 kton, and sometimes much larger, and mostly the tests have been carried out at the Nevada Test Site, although some have been conducted elsewhere, for instance 'Rulison' which took place near Grand Valley, Colorado, in 1969. The Plowshare geothermal concept was described by the American Oil Shale Corporation and the USAEC in 1971 and entails generating power from energy contained in hot, dry rocks.[4] The explosion fractures these as the result of an array of explosives being fired sequentially. Thereafter, water is injected and steam is withdrawn to a surface facility. Figure 5.1 illustrates the sort of plant which could recover energy from a plutonic mass say 3 km deep using hydrogen bombs. Such masses of hot, dry rock as are found in subterranean locations of at least 10 of the western states of the USA could be fractured by such techniques, injected with water and thus turned into important electrical power-generating geothermal sources. The relationships between the precise

number of nuclear explosives necessary, the energy recovered from them
and the energy obtained from the hot rock is obtained from

$$N = \frac{E - NJW}{\varepsilon \tau V_f}$$

where N is the number of nuclear explosives needed, E is the energy needed
to run a 200-MW plant over a time interval of 30 years, i.e.
$1 \cdot 91 \times 10^{11}$ kWh for an 80 % plant factor and a 22 % conversion efficiency,
J is the energy recovered per kilotonne yield, i.e. $1 \cdot 05 \times 10^6$ kWh/kt, W is
the nuclear explosive yield in kilotonnes, ε is the fraction of energy
recovered from rock, assumed to be 0·9, τ is the sensible heat, 180 kWh/m³,
and V_f is the fractured rock volume in cubic metres. Clearly, the amount of
energy necessary to operate such a plant throughout its expected lifetime
must be supplied by the total of the energy derived from the nuclear
explosives and that derived from fractured hot rock. The quantity V_f will be
obtained as the product of the cavity volume and a fracturing-efficiency
coefficient, M. M is itself a relationship expressed by

$$M = \left(\frac{r_f}{r_c}\right) e_h$$

where r_f is the fracture radius, r_c is the cavity radius and e_h is the
enhancement factor. This latter has never been accurately determined, but
there is evidence from experiments in Lithonia granite in which fractured
volumes increased by a factor of 36 when blasting was effected to a free
surface. However, a pre-existing cavity will not act quite as a free surface.[5]
This is because the geometry intercepted is actually limited and the
alteration in density between the cavity in question and the base rock will be
much less than in free-surface blasting.

Experiments have demonstrated that stresses required to widen existing
cracks are much less than those necessary actually to initiate new ones.
Difficulties to be solved in carrying out the concept include finding the
correct array in order to solve the hydraulic-flow problem. The water and
steam must be transmitted through low-permeability regions without
excessive channelling and the dissolution of silica is another factor which
bears close analysis. A fair amount of this may well be transported with the
steam into the turbine system at the high temperatures and pressures
prevailing and, if the quantity becomes too large, an intermediate heat
exchanger could be a necessity at the surface. L. A. Charlot has also
indicated the possibility that, under these conditions, radionuclides
normally resident in the molten rock may also have enough solubility in the

steam to rise to the surface.[6] A restricting factor in application of nuclear techniques may also relate to the size of explosives needed. Larger devices decrease power costs, but there is uncertainty about the limiting sizes which ought to be used. Environmental hazards may arise.

5.2. ENVIRONMENTAL CONSIDERATIONS

The effects of nuclear explosions include ground movement and the emission of radiation.

5.2.1. Ground Movement

When a subterranean nuclear explosion occurs, there is a great release of energy which vaporises the rock and forms an expanding and spherical cavity, the pressure exerted by the latter forcing a shock wave outwards. When this reaches the surface of the ground, it causes vibration to take place and can cause damage over a considerable area. Hence, it is necessary to carry out a full regional survey, perhaps up to 100-km radius from the detonation centre, prior to effecting the explosion. The major danger is the possibility of damaging buildings through such a seismic event of man-made origin. Structural damage will be a consequence dependent upon distance from the detonation point, the geology of the area, the depth of burial of the device, the number and yield of the explosives as well as the natural vibration–response frequency of the individual structures concerned. J. A. Blume has outlined methods for making estimations of the results of explosions and shown that accurate predicting of anticipated structural damage is possible, following an adequate survey. This is very important because even a 100-kt device can cause significant damage at quite large distances and from this, it may be inferred that the use of large nuclear devices, i.e. those in megaton ranges, would appear to be almost everywhere impractical.[7] In the case of the Rulison experiment, for instance, to which reference was made above (section 5.1), a mere 40-kt explosion was involved and yet caused, in a rather isolated region of Colorado, no less than $US90 000 worth of damage to buildings! Actually, nuclear detonations at the Nevada Test Site have also created small-scale aftershocks resulting from minor movements on pre-existing fault planes and due to the release of natural strain energy. In areas of natural seismicity, there is a possibility that such detonations might trigger seismic events sooner than would otherwise be the case. Sandquist and Whan have argued that this might assist in reducing the environmental impact of a natural earthquake because it would promote the liberation of natural strain energy before it accumulates enough to produce such an earthquake.[8]

5.2.2. Emission of Radiation

Table 5.1[4] lists the potentially volatile radioactive nuclides remaining after 180 days have elapsed from the detonation of a 1000-kt nuclear explosion in igneous rock.

At the present time, nuclear explosives, including those relatively clean ones deriving most of their energy from fusion, need a fission trigger which is composed of 3 kt or more of fission energy from uranium-235 or plutonium-239 and provides the high temperature essential to initiate the fusion process. The fissioning of the trigger produces some 200 different isotopes of about 36 elements of atomic masses ranging from about 75 to 160. Mostly, these are radioactive, but short-lifed. Induced radioactive materials are formed as a result of neutron capture in stable elements inside the nuclear device and also in elements in environmental soil materials such as Na, Si, Al, Fe, Mn, etc. The surrounding of the nuclear device with neutron-absorbing materials such as boron will reduce this effect and measures of this sort are certainly desirable where geothermal stimulation is the objective. Probably the most important radiological problem is caused by tritium which enters the ground-water system. As was noted earlier in Chapter 3 (section 3.2.4(iii)), this radioisotope enters the water molecule and therefore travels at the same velocity as ground water. This is the basis of its employment as a tracer and also a 'dating' tool. Since a time lapse of 10 half-lifes for a specific radioisotope produces a reduction in activity by a factor of 1000, and a lapse of 20 half-lifes reduces the activity a million times, even in ground water flowing at 1000 ft/year, tritium activity would be reduced, with a half-life of 12·26 years, about 1000 times within 40 km or so of the site of the detonation. Although tritiated water is a good tracer, it suffers from the difficulty that montmorillonite clays preferentially absorb it, hence introduction of artificial tritiated water is not advisable. Tritium is not too great a biological hazard because of its short biological half-life of 12 days, but if released in large quantities, it can easily become a hazard. Of course, if the nuclear explosion is utilised to stimulate an aquifer, the energy necessary would be lower than for deep, hot, dry rock and therefore the quantity of radioactivity released will also be lower. Another theoretically possible problem is that of the leakage of radiation into the atmosphere from underground nuclear explosions, but where these have been carried out below 4000 ft, this has not occurred and is considered unlikely. The

TABLE 5.1

POTENTIALLY VOLATILE RADIONUCLIDES REMAINING AFTER 180 DAYS HAVE ELAPSED FROM THE DETONATION OF A 1000-kt NUCLEAR EXPLOSION IN IGNEOUS ROCK

Nuclide	Half-life	Fission (kCi)	Fusion[a] (kCi)	Radiation concentration guide[b] (pCi/ml)		Volatility in steam (80 atm, 623 K)
				Air	Water	
Kr-85	10·8 years	20	0·06	0·3	—	Permanent gas
Sr-90	28·8 years	150	0·45	3×10^{-5}	0·3	Gaseous precursor
Ru-103	40 days	1150	3·45	0·003	80	Oxidising, high; reducing, low
Ru-106	1 year	1000	3·0	2×10^{-4}	10	Oxidising, high; reducing, low
Sb-125	2·7 years	60	0·18	9×10^{-4}	100	Apparently high
Te-127	109 days	90	0·27	0·001	50	High
Cs-137	30 years	180	0·54	5×10^{-4}	20	With CO_2, low; without CO_2, high (30%)
			Radioisotopes induced in soil			
H-3	12·26 years	220	20 290[c]	0·2	3 000	Permanent gas and tritiated vapours
Na-22	2·6 years	—	0·6	3×10^{-4}	30	
P-32	14·3 days	2	2·5	0·002	20	—
S-35	88 days	29	40·0	0·009	60	—
Ar-37	35 days	70	200	100	—	Permanent gas
Cs-134	2 years	14	18·3	4×10^{-4}	9	With CO_2, low; without CO_2, high (30%)

[a] These quantities assume linear scaling of 50-kt fission and fusion devices. Assume a nominal 3-kt fission trigger for the fusion device.

[b] Taken from CFR Part 20 for unrestricted areas. They are maximal values not taking into account the solubility of compound.

[c] Assumes 2 g of residual tritium per kilotonne of fusion yield.

equally theoretical possibility of radioactive contamination of ground-water aquifers by seepage is also considered unlikely. The radionuclides most likely to be involved in seepage were it to occur would be the gaseous elements xenon, argon, krypton, iodine and tritium. Even these, however, would have to follow complex paths through fissures in order to pass into ground water or reach the surface of the Earth. Sandquist and Whan believed it unlikely that more than 5 % of total radioactivity would escape even under the worst circumstances.

5.2.3. Additional Effects
One additional effect is the possible surficial extrusion of molten rock as a result of a subterranean nuclear explosion, in other words a volcanic eruption! This will depend upon the proximity of pre-existing magma and also on the availability of open fractures or perhaps older vents through which magma could migrate to the surface. Another is that water near the surface may be close to the temperature at which it will flash into steam. The nuclear detonation might cause a surficial pressure release which could cause flashing of the water and therefore a hydrothermal explosion.

5.3. PROSPECTS

There is little doubt that nuclear stimulation for geothermal steam production is most unlikely to be permissible in populated regions, but, since the effects from explosives of yield less than 100 kt can be fairly well predicted, there is every reason to anticipate the utilisation of the technique in suitable, i.e. remote, ones. The environmental impact of geothermal steam production seems generally to be limited to the area immediately surrounding the zone of application. Nuclear explosives have proved useful for large-volume fracturing in subterranean formations. One interested company has been Battelle-Northwest, acting as a subcontractor to the American Oil Shale Corporation and they have determined that the technique is economically attractive.[9] A related programme has been carried out by the University of California in an attempt to identify problems involved in applying the geothermal steam to desalination of sea water.[9] Here, however, high costs arise in coping with corrosion, scaling and waste disposal of brines. There is also a programme in the USA to stimulate natural-gas production from deep, low-permeability gas fields using the 29-kt Gasbuggy experiment and also the Rulison experiment to which allusion has been made already. In the USSR, nuclear fracturing has

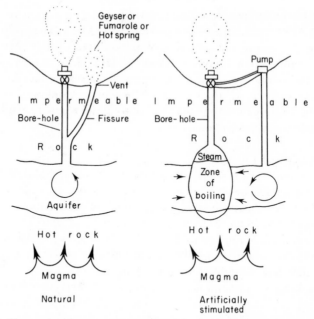

FIG. 5.2. Comparison between a natural and an artificially stimulated geothermal field.

been successfully demonstrated in regard to the stimulation of oil production. As regards geothermal energy, it appears likely that the optimal (low yield) range of nuclear explosions lies within the spectrum of 5–100 kt.

The utilisation of explosive stimulation of geothermal aquifers was proposed by R. Raghavan, H. J. Ramey Jr and P. Kruger in 1971, who suggested that the same approach might be used as in stimulation of hydrocarbon production.[10] This concept would involve the use of low-yield nuclear explosives in order to create a large-diameter well-bore (rubble chimney) in a geothermal aquifer considered capable of replenishing its own produced fluids. The differences between a natural geothermal field and an explosion-stimulated one are illustrated in Fig. 5.2. With careful control of the production rate and steam quality in an explosion-fractured well-bore, it may prove possible to optimise the production of energy from the reservoir.

5.3.1. Corrosion Control
Steam reacts with a number of minerals to form volatile gaseous

compounds which can thereafter be transported and be deposited on piping in steam turbines as well as in the condenser of a nuclear-stimulated geothermal power plant and F. G. Straub has demonstrated that the deposition of salts and silica together with associated radioactivity on turbine blades is dependent upon the impurities in the steam and also upon the steam pressure.[11] As regards salt deposits, these are soluble and can simply be washed off with water.[12] However, silica, as well as iron oxides and calcium carbonate which may also occur, is insoluble in water. Therefore they constitute a very serious problem because they can distort the configuration of turbine blades and thus reduce efficiency, lead to turbine imbalance and cause vibrations which can damage the turbine. Some of the water-insoluble deposits can be removed by washing with sodium hydroxide *if* the deposition rate is not high, but this is a time-wasting operation and must be effected with great care. As well as silica, radionuclides such as ^{125}Sb, ^{137}Cs and ^{134}Cs may also precipitate on turbine blades. Some of these may be avoided by scrubbing the steam with high-purity water before it passes into the turbine. The unfortunate part of this procedure is that it lowers the temperature of the steam and, by degrading its energy, reduces the efficiency of the operation. Additionally, after the scrubbing is completed, the steam is at its saturation point and its use in the turbine could very well lead to erosion of the turbine blades by the impinging on them of water droplets. Ultimately, the steam would have to be reheated.[12] It appears that scrubbing is a feasible means of purifying steam up to temperatures in the neighbourhood of 300 °C, corresponding to saturation pressures of 80–100 atm. A number of salts and hydroxides can be removed by the treatment. O. H. Krikorian has proposed alternative methods. A solid-phase scrubber, e.g. limestone, could be an effective silica-remover.[12] A heat exchanger might be added and water boiled in it, the steam from the nuclear outlet being condensed.

5.3.2. Radioactive Contents of Steam

J. Green and R. M. Lessler noted the radioactive contaminants which might be volatile in steam after 180 days from thermonuclear explosives or fission and the fission products are ^{85}Kr, ^{103}Ru, ^{100}Ru, ^{106}Rh, ^{125}Sb, ^{127}Te and ^{137}Cs, the induced activities being tritium, sodium-22, phosphorus-32, sulphur-35, argon-37 and caesium-134, all being gamma emitters except for ^{106}Ru, tritium, phosphorus-32, sulphur-35 and argon-37.

Thermonuclear explosive has a relatively low level of gamma radiation for volatile radionuclides, but a very high level of beta activity due to tritium. The induced activities are indicated in Table 5.1 and they represent an explosive devoid of neutron shielding. For every 15 cm of boric acid neutron shielding, it has been calculated that there is a factor of about 10 reduction in activation of the subterranean formations.[13] Induced activities also depend upon the nature of the geology, the calculations in Table 5.1 being for Hardhat granite, and the details of the design of the explosive. Non-condensable gases present in steam are separated in the condenser and, in natural steam fields, exhausted from this by venting to the atmosphere. In a nuclear-stimulated geothermal power plant, the gases could be collected and the radioactive contaminants separated or concentrated and subsequently stored underground. If thermonuclear explosives were used, more argon and tritium would be produced and once the latter exchanged with hydrogen in the steam or water, it would be difficult to remove and measures capable of ensuring its safe containment would have to be taken.

REFERENCES

1. Austin, C. F., Austin, W. H. Jr and Leonard, G. W. (1971). *Geothermal Science and Technology: A National Program.* Tech. Ser. 45-029-72, US Naval Weapons Center, China Lake, California.
2. Goldsmith, W., Austin, C. F., Wang, H. C. and Finnegan, S. (1968). Stress waves in igneous rocks. *J. Geophys. Res.*, **71**, 8.
3. Austin, C. F. and Leonard, G. W. (1973). Chemical explosive stimulation of geothermal wells. In: *Geothermal Energy*, ed. Paul Kruger and Carel Otte. Stanford University Press, Stanford, Ca., pp. 269–92.
4. American Oil Shale Corporation–US Atomic Energy Commission (1971). *A Feasibility Study of Geothermal Power Plants.* PNE-1550.
5. Burnham, John B. and Stewart, Donald H. (1973). Recovery of geothermal energy from hot dry rock with nuclear explosions. In: *Geothermal Energy*, ed. Paul Kruger and Carel Otte. Stanford University Press, Stanford, Ca., pp. 223–30.
6. Charlot, L. A. (1971). *Plowshare Geothermal Steam Chemistry.* BNWL-1614.
7. Blume, J. A. (1969). Ground motion effects. In: *Proc. Symp. Public Health Aspects of Peaceful Uses of Nuclear Explosives.* USAEC Tech. Report No. SWRHL-82.
8. Sandquist, G. M. and Whan, Glen A. (1973). Environmental aspects of nuclear stimulation. In: *Geothermal Energy*, ed. Paul Kruger and Carel Otte. Stanford University Press, Stanford, Ca., pp. 293–313.
9. Special Report (1973). *Ground Water and the Geothermal Resource.* Geraghty and Miller Inc., Water Res. Bdg, Manhasset Isle, Port Washington, New York, 14 pp.
10. Raghavan, R., Ramey, H. J. Jr and Kruger, Paul (1971). Calculation of steam extraction from nuclear explosion fractured geothermal aquifers. *Trans. Amer. Nuc. Soc.*, **14**, 695.

11. Straub, F. G. (1964). Steam turbine blade deposits. *University of Illinois, Bull. Eng. Exptal Station Bull.*, *Ser. No. 364*, **43**, 59.
12. Krikorian, O. H. (1973). Corrosion and scaling in nuclear stimulated plants. In: *Geothermal Energy*, ed. Paul Kruger and Carel Otte. Stanford University Press, Stanford, Ca., pp. 315–34.
13. Lessler, R. M. (1970). Reduction of radioactivity produced by nuclear explosives. In: *Symp. Engineering with Nuclear Explosives, Las Vegas, Nevada*, January 14–16, 1970, CONF-700101 (2). Clearinghouse for Federal Sci. & Tech. Information, US Department of Commerce, Springfield, Va.

CHAPTER 6

The Geysers Geothermal Field

The Geysers geothermal field has a power plant which is the largest geothermal installation in the world and, to date, the only geothermal power plant in the USA and it is appropriate here to examine it in some detail. Located on the north bank of Big Sulphur Creek (v. Fig. 6.1) it lies 21 miles northeast of Geyserville and 17 miles east of Cloverdale near the coast of California, west of Sacramento. The existing geothermal facilities are some 90 highway miles north of San Francisco and sited on the steep slopes of a canyon near an extinct volcano, Cobb Mountain. The first recorded sighting of it was by William Bell Elliott, an explorer and surveyor who was looking for grizzly bears in 1847 and came upon steam emerging from a canyon for about a quarter of a mile of its length. By 1975, units in service produced more than 500 000 kW employing steam at about 100 psi and 177 °C piped directly from wells which tap a dry-steam reservoir. Table 6.1[1] shows the plant capacity of The Geysers covering more than a dozen units as of 1976.

It is anticipated that the output will increase greatly in the years ahead and the Electric Power Research Institute at Palo Alto, California, has suggested that as much as 6 MW may be forthcoming from The Geysers by the end of this century. It is interesting to compare this prediction with estimates of the National Petroleum Council in 1972 that the geothermal generating capacity in the USA (primarily in California and Nevada) could reach anything between 7 and 19 MW by 1983. A much more grandiose appraisal was given by the Geothermal Resources Research Conference which stated that, given an adequate research and development programme, no less than 132 000 MW of geothermal power could be generated by 1985 and a fantastic 395 000 MW by the year 2000. It is confusing to examine all these estimates which vary so widely, but they reflect insufficient fundamental data regarding the detection, exploitation

100

TABLE 6.1

THE ELECTRICAL GENERATING CAPACITY OF THE GEYSERS UNITS

Construction date	Unit	Unit capacity (kW)	Cumulative total plant capacity (kW)
1960	1	11 000	11 000
1963	2	13 000	24 000
1967	3	27 000	51 000
1968	4	27 000	78 000
1971	5	53 000	131 000
1971	6	53 000	184 000
1972	7	53 000	237 000
1972	8	53 000	290 000
1973	9	53 000	343 000
1973	10	53 000	396 000
1974	11	106 000	502 000
1976	12	106 000	608 000
1977	13	135 000	743 000
1976	14	110 000	853 000
1977	15	55 000	908 000

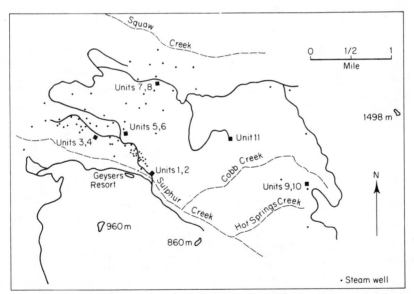

FIG. 6.1. Map of The Geysers, California, USA, showing the layout of the various units north of Sulphur Creek and locations of steam wells. Depth of steam wells given in metres.

and utilisation of geothermal energy and give some idea of how far we have still to go to carry out these activities effectively. However, there is no doubt that The Geysers installations make an impressive beginning for the USA in the tapping of a geothermal energy source.

6.1. GEOLOGY

Figure 6.2 illustrates the general geology of the field. The Geysers comprise a geological formation of fractured shales and basalts of Jurassic–Cretaceous age of high permeability caused by fracturing which resulted from seismic events in the early Cenozoic era. Fault and shear zones which developed also contribute to the permeability and, as a consequence, these and the fractures permit a large reservoir to exist. In this, steam and water are believed to coexist and temperatures range between 260 and 290 °C, the shut-in pressures of wells deeper than a couple

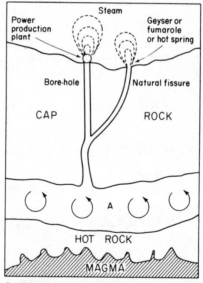

A = Hot water under pressure in aquifer and boiled from below

◯ =Circulation pattern in voids

FIG. 6.2. The Geysers: a natural geothermal field. Cross-section of the subterranean geology in terms of heat and water circulation patterns.

of thousand feet being between 450 and 480 psig. The *total* capacity of the area has not yet been determined. Early on, wells were drilled to depths of between 200 and 1000 ft, but, with improvements in techniques, much greater depths of up to 9000 ft have been attained. Many of these very deep wells are far away from the natural steam outcrops and produce steam flows which, in one case, was as high as 380 000 lb/h. Near the natural steam vents, the early steam wells were adjacently drilled on 200–500-ft centres to depths of between 400 and 1000 ft and these produced steam flows in the range of 40 000–80 000 lb/h.

Geologists believe that a mass of molten magma is still in the process of cooling down some 20 miles below the crust of the Earth, part of it approaching nearer the surface as a result of the fracturation noted above in the overlying rocks. This hot magma is responsible for the natural steam vents (fumaroles) and such steam, deriving from cooling magma, is termed magmatic steam. This is in contrast to meteoritic steam which forms when surface water seeps down into porous rock heated by magma. Meteoritic steam is perhaps the major source of geothermal steam. However, it is not yet clear exactly how the steam at The Geysers is really formed.

6.2. EXPLOITATION OF STEAM

In 1967, the Union Oil Company of California entered into partnership with the original developers of The Geysers, namely the Magma Power Corporation and Thermal Power Company and a great deal of development went on after that time (*v.* Table 6.1). The steam produced has been sold to the Pacific Gas and Electric Company for use in condensing turbines driving electric power generators. The reservoir pressures noted above in section 6.1, of course, reduce to about 125 psig at the wellhead as a result of the incoming steam expanding and cooling; consequently, the turbines are designed to operate at 80–100-psig intake pressure. Generating facilities must be constructed close to the wells because geothermal steam rapidly loses heat energy when it is transported through a pipeline. Also, in order that they may operate at optimal capacity, the turbines need a constant throughput of steam.

The total capacity of The Geysers, as of any geothermal field, may be assessed as a result of exploration, and appropriate techniques can identify the limits of potentially productive regions. The problem is that this is done only in a very generalised way. For instance, L. E. Garrison described the known gravity anomaly at The Geysers as covering 100 square miles and within this, there are several scattered wells drilled to great depth *which did not produce steam!*[2] This is probably related in some way to inadequate permeability; indeed, it is hard to think of any other explanation. But this could not be predicted from the gravimetric measurements. What this means is that only data gathered directly from drilling activities can give the complete picture of any geothermal field's capacity. Drilling faces problems in geothermal fields and, at The Geysers, they are connected with the fact that the actual producing zone is one of graywacke sandstone (Franciscan formation), highly metamorphosed and therefore hard to penetrate. Mud drilling causes lost circulation to permeable zones with fluid pressures below hydrostatic. Air drilling improves penetration rates and also eliminated the lost circulation question, but can only be effected below segments where significant water-bearing beds have been cased (otherwise the water would enter the bore-hole and promote cave-in of the formations). There is heavy wear on equipment due not only to abrasion but also to heat. When steam is encountered, air circulation pressures rise and annular velocities approach sonic velocity which causes tool joints to take the wear of high-speed particle impact.[3]

The production rate attainable from a completed steam well is sensitive to sizes of both hole and casing. Such wells are capable of very high mass-flow rates which may be limited by the frictional resistance of both of these. Research shows that a steam well at The Geysers conforms to the classic expression for the performance of a gas well, i.e.

$$W = C(P_s^2 - P_f^2)^n$$

where W is the steam flow rate in lb/h, C is the factor which is a function of time, reservoir matrix, fluid properties, flow rate, wellbore conditions, P_s is the static reservoir pressure in psia, P_f is the wellbore pressure at steam entry in psia and n is an exponent, usually constant with W within the range of usual production rates, $0.5 \leq n \leq 1.0$.[4,5] Production, W, is dependent, therefore, upon the bottom hole pressure, P_f, i.e. the sum of the wellhead flowing pressure, the total frictional resistance pressure and the column weight of the fluid. The cost per foot of drilling has been found to increase sharply with depth at The Geysers, so that the chance of locating an *economic* steam-producing well gets less with increasing depth as a result of the fact that the length of the steam-flow conduit increases also. There is no certain way in which the productivity of a well can be predicted before it is drilled, hence the values for C and n also cannot be accurately predicted. Calculations can be made, however, using typical values from typical existing wells.

6.3. THE PERFORMANCE OF THE RESERVOIR

Interested parties have studied the reservoir of The Geysers vapour-dominated system for many years. The field performance has demonstrated that a constant rate of production cannot be sustained for ever; individual wells have declined and extra wells have been necessary in order to maintain the supply of steam to the generating units. Incidentally, this has also been the case at Larderello. Analysis of two wells at The Geysers, part of a block-of-power which has been producing steam for generating unit 3 since 1967, showed that, throughout their producing life, the wellhead flowing pressure has been more or less constant.[3] It may be inferred that the variation and decline in productivity are not the result of changing surface conditions. Pressure build-up tests on the two wells in question showed that no proximate-wellbore alteration in permeability occurred. This leads to the conclusion that there has been no plugging of the wellbore or the formation. The overall implication is that pressure depletion in the reservoir itself is a factor in the production decline.

6.4. INSTALLATIONS AT THE SURFACE

Surface installations are for collecting the steam from several wells and delivering it to a generating plant in a condition of purity, i.e. free of particles and moisture. It is also desirable to achieve this operation with the minimal loss of energy. Hence, insulation is necessary, but this must offer minimal resistance to flow. Separating devices and vessels must be available for removal of liquid and solid particles and there must also be adequate safety features which can relieve, if necessary, line pressure if a plant shutdown occurs. The facility must be so designed that steam is not trapped, cooled and condensed *en route*. This is because if condensate forms and drips back into the line, it will further deplete the energy of the steam. A system was developed at The Geysers in order to deliver 2 million lb/h of dry, superheated steam through two separate systems to two units jointly generating 110 MW of electricity.[3]

Near each wellhead is a meter which continuously records the production rate and producing pressure of the well. Routine measurements of enthalpy and steam quality are also carried out.

Steam at The Geysers contains about 1 % of non-condensable gases in roughly the amounts given in Table 6.2.[1]

As noted above, the steam also contains impurities, including dust which

TABLE 6.2

NON-CONDENSABLE GASES CONTENT OF
STEAM AT THE GEYSERS (TOTAL QUANTITY
IS 1% AND AMOUNTS OF THE VARIOUS
CONSTITUENTS CONTRIBUTING TO THIS
ARE GIVEN)

Non-condensable gas	Quantity (%)
CO_2	0·79
NH_3	0·07
CH_4	0·05
H_2S	0·05
N_2, Ar	0·03
H_2	0·01
	1·00

may build up on the inside of the turbine blade shrouds, a phenomenon which may cause failure. This has been a problem mitigated by the installation of heavy-duty replacement blades and shrouds in earlier units.[1]

6.4.1. The Construction Materials

Studies were effected to assess the suitability of various materials for the mechanical equipment and piping prior to the commencement of the detailed designing of the first unit. The steam emerging from the wells with a small quantity of superheat was found to be relatively non-corrosive and in this part of the operation, carbon steel can be employed for piping. Also, the turbines do not need any special corrosion-resistant materials. However, as the steam condenses, non-condensable gases become more concentrated and H_2S partly oxidises to form dilute sulphuric acid. Consequently, the steam and condensate become much more corrosive and carbon steel is inadequate as also are copper-based alloy, cadmium and zinc. Austenitic stainless steels or Al or epoxy-fibre glass are quite suitable.

Hydrogen sulphide in the atmosphere may cause difficulties with the electrical equipment, again because of corrosion. The lack of appropriate components in unit 1 caused problems with electrical contacts and other metal parts. Aluminium appears to be very satisfactorily resistant, just as much so as stainless steel and some precious metals such as platinum.

Protective relays are especially susceptible to corrosion and special ones made of non-corrosive materials were supplied for units 2, 3 and 4. From unit 5, relays, communication equipment, 480-V switchgear and generator excitation cubicles were placed in a clean environment comprising three rooms on three levels maintained at a slight positive pressure using clean air from activated carbon filters.

6.5. OPERATION

The initial investigations were aimed at establishing the sufficiency of the steam source; at that time, steam suppliers had to demonstrate that there was a sufficient flow from existing wells to supply the first turbine generator unit. The steam has a constant enthalpy of 1200–1205 Btu/lb. To fully exploit the steam energy, it was considered necessary to use condensing steam turbines exhausting below atmospheric pressure. There was no source of condenser cooling water in the region and therefore cooling towers were incorporated into the cycle as a heat sink.

6.5.1. Cooling Towers
The cooling towers are of the induced-draft type and the structural supports are of redwood. The tower siding is transite. Polyvinyl is favoured as fill material because of its fire-retarding qualities. Cooling-tower basins are made of reinforced concrete coated with coal-tar epoxy in order to prevent deterioration of the concrete. The cooling towers are designed to cool the condensate from 48 °C to 27 °C at 18 °C wet bulb.

6.5.2. Turbines
The turbines are manufactured mainly of manufacturers' standard materials for low-pressure and low-temperature service. Typically, the blades and nozzles are of 11–13 % chrome steel, carbon steel being utilised for the casings. Austenitic stainless steel inserts are provided in casings opposite the rotating blades in order to prevent moisture erosion of the casings. The steam inlet conditions for the turbine-generator units are shown in Table 6.3.[1]

In the case of the earlier units, lower steam pressures were employed as a result of the fact that their supply of steam originated from the shallow, lower-pressure reservoir. Later units are supplied from the deep, higher-pressure reservoir.

6.5.3. Electric Generators
The electric generators used at The Geysers are quite typical of similarly

TABLE 6.3

STEAM INLET CONDITIONS FOR THE TURBINE-GENERATOR UNITS

Unit	Rating (kW)	Steam flow (lb/h)	Steam pressure (psig)	Temperature (°C)
No. 1	12 500	240 000	93·9	175·5
No. 2	13 750	255 475	78·9	172·2
No. 3 and No. 4	27 500	509 600	78·0	172·2
Nos. 5–10	55 000	907 530	113·7	179·4
No. 11	110 000	1 808 000	113·7	179·4

sized units employed elsewhere, the larger ones being hydrogen-cooled and automatically purged under certain conditions of difficulty. Outdoor generator oil circuit breakers are utilised. The potential transformers are also outdoors and their potential taps to the bus are supplied by the cable-bus manufacturer.

Four aluminium cables are utilised per phase for each generator, giving a rating of 3000 amperes (A). Cables from each generator oil circuit breaker are directly connected to the single main transformer terminals and are shielded with cross-linked polyethylene insulation.

On units 1 to 4, three different types of excitation system are employed. Thereafter, static excitation is utilised with power potential transformers and saturable current transformers. The eliminating of commutators is desirable in the environment of The Geysers.

6.5.4. The Main Transformers

Units 1 to 4 each possesses its own step-up transformer, essential in these early units which contributed to the local power supply. Subsequent increase of the unit size to 55 000 kW was accompanied by the utilisation of one main three-phase transformer of 132 000 VA for each two units. This was an economic measure aimed at the reduction of costs. The transmission voltages have ranged from 60 000 V to 230 000 V. For units 3 to 10, dual voltage step-up transformers were acquired, but, from unit 11 onwards, single voltage 230 000-V transformers are feasible because the combined plant outputs necessitate 230 000-V operation.[1]

6.5.5. Metering

The steam purchase contract stipulates payment for power delivered to transmission. Units 1 and 2 and units 3 and 4 have transmission voltage metering sets of 60 000 V and 115 000 V, respectively. As 230 000-V

metering equipment is expensive, units 5 and 6 and subsequent units are
metered at the low-voltage side of the main transformer. The transformer
losses are determined from ampere-squared-hour and volt-squared-hour
meters and subtracted.[1]

6.5.6. The Power Cycle

Steam from wells is introduced into the turbines thus exhausting the direct
contact condensers below them and the combined condensed steam and
cooling water are pumped by two condensate pumps to the cooling water.

The turbine back pressure on all units is approximately 100 mm Hg
absolute. Cooled water from the tower basin is returned back to the
condenser under gravity and the vacuum head developed by the latter.

The evaporation rate of the cooling tower is less than the turbine steam
flow and in consequence, an excess of water develops. The flow depends
upon the dry-bulb temperature and the relative humidity, but under all
conditions of operation, an excess persists. Over a period of years, this has
been returned to the steam suppliers for reinjection through wells into the

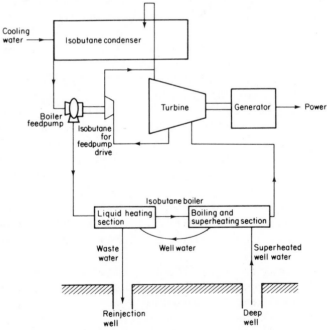

FIG. 6.3.   The vapour-turbine cycle.

steam reservoir. It was at first thought that this process might quench the producing steam wells, but no deleterious effects appear to have taken place. Actually, the reinjection may very well prolong the productive life of the steam reservoirs because there is a possibility that there is more heat in the reservoir than there is vapour available to extract it.

Two-stage steam-jet ejectors are utilised in order to purge the non-condensable gases from the turbine condensers. The condensers for the ejectors are again of direct contact design.

A simplified diagram of the vapour-turbine cycle is shown in Fig. 6.3. Here, isobutane is shown as the fluid utilised, but others may also be employed. J. H. Anderson stated that this one, however, has demonstrated the optimal economics for developing power from water at a temperature of about 180 °C. Another possible fluid which has been used in the USSR is freon.[6,7]

6.6. COSTS

It has been determined that the economics of operating a geothermal power plant are comparable to those incurred in running other electrical generating stations. The fixed charges of a conventional fossil-fuel plant are similar to those for a geothermal power plant. Either of these costs much less than a hydropower plant or a nuclear plant. As regards other expenditures such as maintenance, these are about the same for all. It is very important to remember the low pollution hazards of geothermal power plants, however. Direct comparisons of costs between energy generation systems are not really feasible and very hard to make, of course, as a result of the practically continuous increasing in costs of conventional fuels.

Table 6.4[8] shows a 12-year average of geothermal costs taken over the period 1961 to 1972 inclusive at The Geysers, costs which will vary much less than for other fuels because the only variable involved is the input of steam, other prices being almost constant (this is because the number of personnel required is small, therefore wage rises are not significant).

Two factors are significant in geothermal cost consistency and these are

1.   need for increased profit in order to promote new exploration and field development as well as expansion of existing facilities;
2.   increase in efficiency of a geothermal power plant by utilisation of a multistage set-up using as near as possible to *all* of the aspects of a geothermal system.

TABLE 6.4

AVERAGE GEOTHERMAL COSTS AT THE GEYSERS FROM
1961 TO 1972 INCLUSIVE (CAPACITY FACTOR = 77·5%)

Items of expenditure	Power cost (mills/net kWh)	Annual cost ($US/kW)
Fixed charges	2·25	14·08
Maintenance	0·387	2·43
Operation	0·208	1·30
Steam	2·587	—
Total	5·432	17·81

Unfortunately, some costs of geothermal power plants are governed by factors over which Man has little control, for instance the geological setting. There are other factors over which we have no control at all, including the pressure, temperature and fluid characteristics of a geothermal reservoir which have developed over millions of years in some cases.

It is probably good for our collective soul to be reminded thus that it is not we, but, as the Psalmist says (107: 35), 'He' who 'turneth the wilderness into a standing water and dry ground into watersprings'!

REFERENCES

1. Bolton, R. S., Bowen, R. G., Groh, E. A. and Lindal, Baldur (1977). Geothermal energy technology. Section 7 in: *Energy Technology Handbook*, ed. Douglas M. Considine. McGraw-Hill Book Company, New York, pp. 1–57.
2. Garrison, L. E. (1972). Geothermal steam in the Geysers–Clear Lake region, California. *Geol. Soc. Amer., Bull.*, **83**, 1449–68.
3. Budd, C. F., Jr (1973). Steam production at The Geysers geothermal field. In: *Geothermal Energy*, ed. Paul Kruger and Carel Otte. Stanford University Press, Stanford, Ca., pp. 129–44.
4. Craft, B. C. and Hawkins, M. F., Jr (1959). *Applied Petroleum Reservoir Engineering*. Prentice-Hall, New York, pp. 326–27.
5. Cullender, M. H. (1955). The isochronal performance method of determining the flow characteristics of gas wells. *AIME*, **204**, 137.
6. Anderson, J. H. (1973). Vapour turbine cycle for geothermal power generation. In: *Geothermal Energy*, ed. Paul Kruger and Carel Otte. Stanford University Press, Stanford, Ca., pp. 162–75.
7. Moskvicheva, V. N. (1971). *Geopower Plant on the Paratun'ka River, USSR*. Academy of Sciences.
8. Cheremisinoff, Paul N. and Morresi, Angelo C. (1976). *Geothermal Energy Technology Assessment*. Technomic Publishing Co. Inc., Westport, Conn., 164 pp.

CHAPTER 7

The Geothermal Resources of New Zealand

New Zealand enjoys an enormous geothermal resource, both actual and potential, a thermal region extending along a great belt 50 km wide and some 250 km long located in the North Island between the White Island Volcano in the Bay of Plenty and the volcanic mountains which occur centrally. Currently, exploitation activities are proceeding at Wairakei, Kawerau and Rotorua and a new field is being brought into the picture at Broadlands. All lie within the volcanic area of Taupo (Fig. 7.1). It will be appropriate to examine Wairakei initially (v. Fig. 7.2) because this is the second major geothermal power station to be constructed and it first supplied electricity in November 1958.

7.1. WAIRAKEI

The New Zealand Electricity Department controls the Wairakei stations and is responsible for the distribution of electricity through the grid system (v. Fig. 7.3). It installed the turboalternators and other equipment. The administration of the actual construction work and subsequent downhole maintenance as well as the designing and installing of some large modifications to the steam transmission system are effected by the Ministry of Works and Development. This body also investigates other geothermal areas, looks into alternative uses for the energy and, with the Department of Scientific and Industrial Research, improves old techniques and devises new ones. There is a scientific team and laboratory with technicians performing fundamental research in both volcanism and geothermal energy at Wairakei.[1]

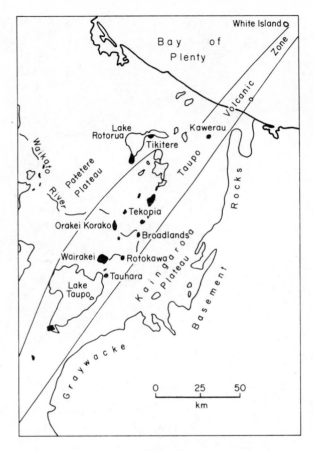

FIG. 7.1.   The Taupo Volcanic Zone, New Zealand.

7.1.1. Geology

Surficial formations comprise pumice and loosely consolidated breccia to a depth of about 125 m and these are underlaid by the Huka Falls Formation, a set of mudstone and siltstone layers sometimes interbedded with breccias and ranging from about 125 m to 200 m in thickness (Fig. 7.4). These latter are usually impermeable and therefore the Huka Falls Formation constitutes an excellent cap rock. Natural activity has been recorded, however, and this demonstrates that there is a certain amount of fissure permeability.

Beneath those mentioned above is the Waiora Formation, reasonably

FIG. 7.2.   General aerial view of Wairakei. This photograph was taken in 1965 and it must be borne in mind that there have been subsequent changes. Reproduced by courtesy of The High Commissioner for New Zealand, London.

well-consolidated breccias with mudstone–siltstone stringer and silicified lenses. To the west of the region, the formation is approximately 650 m deep, but on the eastern side, it has not been drilled to its full depth and must be much deeper. The Waiora Formation is mostly permeable and provides quite a large proportion of the production.

    Underlying the Waiora Formation is the Wairakei Ignimbrite Group and the contact between the two is broken and constitutes the main

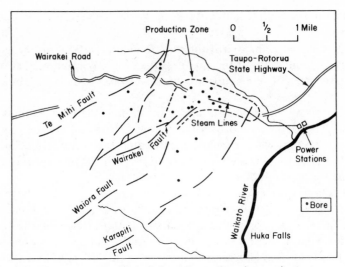

FIG. 7.3. The Wairakei geothermal project region.

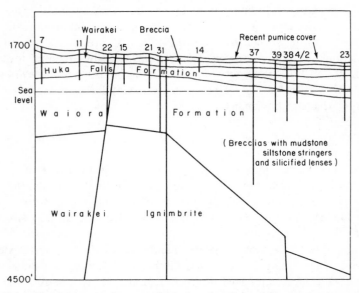

FIG. 7.4.   Northwest–southeast geological section through the production zone of
the Wairakei geothermal project region. Numbers are well numbers.

production zone, the ignimbrite itself having a very low production capacity.

In the outer area, many rhyolites are found in the Waiora Formation, there is a thinning out of this and the Huka Falls Formation and, in the lower levels, the above-mentioned contact is displaced downwards by about 500 m.

There is very extensive faulting in the region, the main fault system following a southwest–northeast direction. There is an occurrence of transverse faulting where the most intense natural activity is found. Drilling has established that the faults can be open at depth and must exert a strong influence upon the subterranean flow conditions.

The field has been exploited now for two decades and there have been marked changes in the underground conditions. At first, the aquifer was full of water with temperature–pressure conditions following the boiling point for a depth relationship until a temperature of 260 °C was attained. Now, there has been an almost uniform pressure decline of more than 21 bars affecting an area much larger than that of production wells, an area roughly delimited by the Wairakei Stream in the north and northeast, the Waikato River to Huka Falls in the southeast, the Waipouwerawera Stream in the south and Poihipi Road in the west. The relationship between the rate of pressure decline and the rate of drawoff is not linear. Thus, an increase in the latter is not at once followed by stabilisation; there is nevertheless an immediate pressure fall. A prolonged period of substantially constant drawoff demonstrates pressures tending towards stable value.[1] Bolton suggested that conditions are being influenced by an inflow, which is supported by the rise in pressure at the beginning of 1968 in a period when the drawoff in the field was actually reduced to only one-third of normal.[1]

The temperatures in the upper levels show a similar trend but at greater depths, no changes are observed.

The trends in individual well outputs are affected by these factors, but the extrapolation of trends in the field's total discharge indicates that the flow tends towards a stable value enough to maintain an output of between 125 and 140 MW indefinitely. For comparison, the output in 1974 was 150 MW.

Some of the wells show an increase and others a decrease in the steam content. The enthalpy of the total discharge is showing a tendency to diminish and Bolton inferred that, although it is always possible that the field will run dry, this will take a long time and, to 1975, there are no indications of this actually happening.[1] At that time, Wairakei produced

FIG. 7.5. General view of Wairakei. Reproduced by courtesy of The High Commissioner for New Zealand, London.

about one-tenth of all energy required by the North Island operating at an annual load factor of 89 %, but this contribution is going to go down as the capacity of the country's electric system goes up.

7.1.2. Wells (v. Fig. 7.5)

The loads imposed by the drilling rigs are carried by reinforced concrete collars at the wellheads and the ground around these is consolidated by grouting in order to render it impervious and supply protection against the upward movement of steam and hot water to the surface around the wells. They also act as safety measures. The drilling is effected using rotary-type drills and rapid cooling facilities have to be provided for the drilling mud. All casings are cemented to the surface (except for the liner). The early wells were of 10-cm and 15-cm diameter and were drilled with rigs of 250-m and 500-m nominal capacity. Mostly, however, the production wells are of 20-cm diameter and are drilled with a rig having a normal capacity of 900 m. Some of these wells penetrate to 1400 m. To last year, 102 wells have been drilled at Wairakei and 68 of these have supplied the powerhouse; the rest comprise investigation wells unsuitable for production and include one drilled to 2400 m using a 3000-m capacity rig.

7.1.3. The Collection and Transmission of Steam

It was originally planned to separate steam and water at each wellhead by using two separate supply systems, one operating at c. 13·5-atm gauge and the other at c. 5·5-atm gauge. The latter (intermediate pressure) steam was intended to provide energy to a heavy-water plant, the high-pressure steam being intended to generate electricity. The heavy-water scheme was dropped, however, but the two-pressure transmission system was retained because tenders had already been called for construction.

A pilot scheme was designed and commissioned to test the feasibility of transmitting the water to the powerhouse where it was to be flashed in order to provide extra intermediate-pressure steam as well as low-pressure pass-in steam at atmospheric pressure. The scheme was satisfactory, but it was begun at a time of accelerated drilling and development of the field and because of a marked increase in total drawoff, the water output declined to a point where the project had to be abandoned—because of water shortage.

Separated steam with the energy of the hot water wasted continued to be the basis for the development of Wairakei until 1974, when measures were taken to enable the latter to be used.

Original working pressures were determined by prevailing circumstances. The ill-fated heavy-water plant was to have received geothermal steam at 3·5 atm and give vapour at about 0·1-atm gauge, requirements necessitating two pressure systems within the plant area. Condensing turbines were envisaged for low-pressure steam absorbing and also topping sets to extract power from the high-pressure wells before exhausting into the distillation plant. The steam from the intermediate-pressure wells could be delivered to the station at approximately 3·5-atm gauge in order to supplement the exhaust steam from the topping sets. The admission pressure for the latter entailed compromise and a figure of 12-atm gauge was arrived at as a sort of guesstimate. This pressure had to be reduced, ultimately to 8-atm gauge, because of the later exploitation of the field.

Modification of the system has been alluded to above in order to make use of heat in the water which was previously wasted. Another pressure system was introduced, therefore, and this was termed the intermediate low-pressure system, operating at 1·67-atm gauge. The result is that four pressure systems are now in use at Wairakei. These are

1.   high-pressure system,
2.   intermediate-pressure system,
3.   intermediate low-pressure system,
4.   low-pressure system.

7.1.4. Wellhead Equipment
Wellhead equipment is primarily intended to separate steam from water as

soon as the mixture emerges from the well and two ways of doing this have been employed. The first effects the separation in two stages, the mixture being conducted round an inverted U-bend above the well, water being thrown off by centrifugal force against the outer wall of this and the steam being drawn off from the inner wall just after the bend; 80–90 % of the water is removed. The wet steam is then conveyed to a top outlet cyclone separator from which 90 % dry steam emerges. A simpler approach is to utilise a Webre-type cyclone omitting the U-bend and here the steam and water mixture is conducted directly to a vertical cylinder through a spiral inlet from the well. In the cylinder, there is a basally located central steam take-off pipe. A ball-float valve is fitted into this in order to prevent water from being carried over into the steam pipelines should accidental flooding occur.

Separated water flows to a drum. At each well, there is a bypass connection which acts to enable the well fluid to be blown to waste when not needed or when the wellhead equipment is out of service for inspection and maintenance. A silencer is provided at each well (Fig. 7.6), also, to receive rejected or bypassed fluids and this comprises two cylindrical concrete vessels joined tangentially to each other and open at the top. The mixture of water and steam impinges on to a cusped steel erosion plate let into the walls of the concrete cylinders at the tangential contact point. The water swirls round the walls and is led to waste after dissipation of the bulk of its kinetic energy through friction and turbulence, the steam escaping from the tops of the cylinders. In consequence, most of the noise is conveyed upwards.

At all the wellheads, flowmeters are placed in the steam take-off pipes and pressure gauges are installed to register the wellhead pressures and also the pressure drops across the equipment. By routine logging of these, a continuous check on the behaviour of the bores may be made.

Dry steam is conducted from the wellhead separators through branch pipes to the main steam-transmission pipelines. The branch pipes have bores varying from 15 cm to 30 cm according to length and well output and the main pipelines are 50, 75, 105 or 120 cm (nominal).

Initially, 50-cm nominal bore pipe was usually the most suitable and in any case, at that time, it was the largest size of seamless pipe available easily. Seamless pipe with no welds was considered necessary. In the second stage, larger pipes were required and 75-cm, longitudinally welded pipes were employed. Subsequently, welded pipes of 105-cm and 120-cm diameters were utilised. These were suitable for the large specific volumes associated with the intermediate low-pressure system. For the high-pressure pipelines, capacity was restricted by pressure-drop considerations. For a total pipeline drop of

FIG. 7.6. Vertical twin tower silencers forming part of the standard equipment at each wellhead (Wairakei). Reproduced by courtesy of The High Commissioner for New Zealand, London.

1·3 atm or so, the 50-cm pipe will deliver 110 tons hourly and the 75-cm pipe about 330 tons hourly. However, the capacity of the lower pressure mains is limited by considerations of velocity. For categories (2) and (3), it was thought inadvisable to exceed 45 m/s owing to the possibility that erosion from slightly wet steam might occur. On this basis, the capacity of the 50-cm, intermediate-pressure line is approximately 90 tons/h and that of the 75-cm pipeline is 206 tons/h. The 120-cm, intermediate low-pressure line operating at about 1-atm gauge just upstream of the letdown valve of the power station has a capacity of 150 tons/h.

In order to take up expansion in the main pipelines, flexible loops are provided at intervals of approximately 300 m or less and these are placed in a vertical plane to allow road access beneath them. They impart practically negligible thrusts on anchor blocks. Each one is capable of absorbing approximately 75 cm of pipe movement and incorporates three hinged angular bellows compensators (resembling a three-pin arch). To obviate risks from seismic shocks and lateral winds, the loops are supported by light steel structures. Between anchors, pipe movement is taken up by means of rollers. On branch lines, expansion is taken up by means of solid bends and loops without the aid of flexible bellows pieces.

En route from the bores to the stations, some steam condenses and the condensate is collected in drain pots formed in the mains at intervals of approximately 150 m, thereafter being discharged to waste through traps. Heat is lost, but some pipeline condensation is beneficial because, through dilution and partial removal several times repeated, the water phase becomes purified in transit. The pipeline acts, in effect, as a scrubber.

Safety valves and bursting discs are provided at every wellhead and flash unit in order to protect the equipment against any abnormal rises in pressure. The main problem in transmitting hot water is to prevent boiling in the pipeline. This was solved by raising the pressure by pumping and lowering the vapour pressure by cooling. This latter was effected by mixing the water from the intermediate-pressure and high-pressure bores.

Geothermal fluids contain chemical impurities and, to resist these, special materials were employed. In the absence of oxygen, mild steel was found to be quite resistant to geothermal fluids and may be utilised for wellhead equipment, steam and hot-water pipes and flash vessels.

7.1.5. The Power Plant (*v.* Fig. 7.7)

At A station, there is an arrangement of three groups of turbo sets operating in series. B station has three mixed-pressure pass-in sets which indicate the design which would have been adopted before had the heavy-water plant

FIG. 7.7.   View of the power house showing low-pressure pipework above and high-pressure/intermediate-pressure pipework below. Reproduced by courtesy of The High Commissioner for New Zealand, London.

not been included in the original plan. In order to ensure, as far as feasible, an undisturbed balance of steam flows when any set in the A station is taken out of service, bypass reducing valves are connected across the back-pressure sets and a dump condenser is provided as a flow substitute for a low-pressure condensing set. If two or more low-pressure sets should be out of service at the same time, it would be necessary to curtail the load on the lower-pressure sets.

Jet condensers are used for the low-pressure sets and the condensed steam is passed to the river with the cooling water. The condensers are freed from non-condensable gases in the A station and the B station by utilising steam-operated ejectors.

The exhaust steam from each back-pressure set would be of the order of 6% wet with dry steam at the stop valve. Separators are incorporated in the pipework system to take out most of the wetness and thus ensure that steam entering the next downstream turbine may be as dry as possible.

As regards the generator sets, the arrangement of three classes of turbines in the A station was influenced by the proposed heavy-water plant and is too cumbersome in consequence, an error not repeated with the B station.

On the cooling system, a riverside pumping house was constructed in order to provide the necessary water. Three years after Wairakei was completed, the Aratiatia hydroelectric scheme was constructed about 3 miles downstream on the Waikato River and raised the level of the river by 3 m. To cater for the two conditions of river level, two points on the characteristic curve were specified when tenders for the pumps were solicited. For the A station, four centrifugal pumps exist for supplying the condensers of the low-pressure sets. Cooling water is passed through rotary strainers before entering the jet condensers and accumulated solid matter is removed by scrapers and flushed to waste. After passing through the condensers, the warmed water enters an underground culvert under the foundation raft and is conducted to an outfall on the bank of the river some 200 m downstream of the pumphouse. The mean river-water temperature is 15 °C (maximum is 21 °C) and the outlet from the condensers has a mean temperature of 29·5 °C (maximum 34 °C).

Turning now to the electrical equipment, the step-up transformers are in single-phase units with one spare unit of each size and the 200-kV substation is of the ring bus–bar type, bars being mounted on post insulators themselves placed on concrete stools. Outgoing 220-kV feeders are taken from a subsidiary 11-kV switchboard itself supplied from the main 11-kV boards through 1:1 transformers which limit the fault level and give voltage regulation facilities.

If an orthodox thermal is compared, this has more auxiliaries. The geothermal one requires only cooling water pumps. All auxiliaries to the A station are fed from 3·3-kV and 400-V station boards and in the B station, each alternator has both a unit transformer and a unit auxiliary transformer which feed a cooling water pump at 3·3 kV and minor auxiliaries at 400 V. Also in the B station, there is a 3·3-kV and a 400-V general-purpose switchboard arranged for alternative feeds. In the A station, there is a 63-kW automatic starting a.c. diesel generator which is intended to supply selected sections of main lighting should an emergency arise.

Neutral points of the 220-kV transformer windings are solidly grounded and those of the alternators also, in the latter case, through voltage transformers which possess alarm relays connected to their secondary windings. Naturally, the 3·3-kV systems in both stations and the A station's 11-kV distribution system are also solidly grounded. The 400-V supplies in both stations possess a multiple-grounded neutral system.

Alternators and step-up transformers are equipped with overcurrent and balanced current protection and all other transformers possess overload and ground leakage protection. All the transformers, with the exception of 400 V, have gas relay protection. Electronic regulators take care of the generation voltage of every set.

Between the A and B stations, there is an auxiliary building housing the compressors for the 220-kV air blast switchgear, the filtration plant for the transformer oil and the 63-kW diesel generator.

7.1.6. Safety Considerations

Safety considerations are stringently observed in that practically all possible contingencies have been allowed for, particularly the necessity to allow for relief of pressure if the total inflow of steam becomes excessive (spring-loaded safety valves are present) and also the necessity of ensuring that a slug of water does not reach the turbines (ball valves prevent large amounts of water from entering the pipelines and there are traps along the mains to deal with condensation plus water detectors to transmit distress signals to the station). It must be noted that the power station operates in parallel with several hydrostations which are so governed that allowance is made for a momentary rise in frequency up to a level of 35 % above normal in the event that load loss occurs. The turboalternators at Wairakei are designed for a maximal overspeed of 15 % and so it is necessary to guard against the Wairakei sets being motored by the hydro sets if the entire group of stations were to lose load. This is done by supplying high-speed, frequency-sensitive relays capable of cutting the electrical connection between Wairakei and the hydrostations by operating the step-up transformer circuit breakers should the frequency rise to 7·5 % over normal.

7.1.7. Efficiency of the Operation

To ensure maximal value from Wairakei, the stations must be operated at the greatest possible load and at high load factor, with absolutely steady conditions. Load dispatching in the North Island is so manipulated that variations in load are as far as possible taken up by the hydro and fuel-fired stations. The Wairakei control engineer endeavours to maintain steam pressures at the stations at values facilitating operation of the sets at the highest possible outputs.

Since commissioning of the station, it has operated at an average load factor of 89 %.

7.1.8. Environmental Effects

Since the Waikato River is used both as a heat sink and to dispose of waste water, it is to be expected that some adverse effects may result. The mean flow of the river is about $125\,m^3/s$ and the average temperature rise from waste heat is $1.5\,^{\circ}C$. A good point is the low content of total dissolved salts in the geothermal fluids, a mere 4400 ppm, since this means that the mean salinity of the river is only increased by a quantity under 40 ppm. To the effects of the geothermal plant, however, must be added those of the hydroelectric dams as well as those due to land reclamation which has led to continuing utilisation of fertiliser.

Inside the field region, the principal adverse effects may be ascribed to the gas content of the geothermal fluids, the consequences of uncontrolled blowouts and subsidence. The gas is almost all CO_2 accompanied by about 5% of H_2S and when this is eventually vented into the atmosphere, some may descend from the powerhouse stack to ground level in bad weather.

As regards blowouts, two have taken place at Wairakei. One was caused by the failure of a well casing after some years of service and the other took place during the drilling of a well. Environmental results include a sudden, increased discharge of heat and water locally and also the possibility of the endangering of other structures in the vicinity. If the blowout is in a remote area, it may be left unattended as was the case with one of the two mentioned above. After 14 years, this sealed itself off and, in the meanwhile, constituted a tourist attraction!

Subsidence at Wairakei has caused an elliptical depression to develop, the maximal rate of subsidence being just under $0.5\,m$ annually, the total maximal subsidence being estimated as a little over $3\,m$. The centre of subsidence is situated about 450 m from the nearest well and 1900 m from the region of maximal drawoff. Unfortunately, the geology of the subsidence zone is not well known.

Ground movement has affected steam mains and drainage channels, but has not seriously affected the field's operation.

Naturally, environmental effects are not always deleterious and against those which are, we must remember the enormous benefits accruing to New Zealand's economy from the existence of this low-cost and reliable source of indigenous energy and the jobs connected with its exploitation. The geothermal plant itself constitutes to some degree a tourist attraction also.

7.2. BROADLANDS, A GAS-DOMINATED GEOTHERMAL FIELD

Like Wairakei, the Broadlands geothermal field lies within the North Island's Taupo Volcanic Zone (v. Fig. 7.8), actually about 25 km northeast of Wairakei. A regional de-resistivity survey defined the area as suitable for exploration and also demonstrated that the field may well be larger than the surface manifestations might have led one to suppose and also that its lateral boundary at depths near 0·75 km is abrupt.[2]

Silica concentrations and Na/K ratios derived from discharge waters suggested a subterranean temperature probably around 260 °C. The

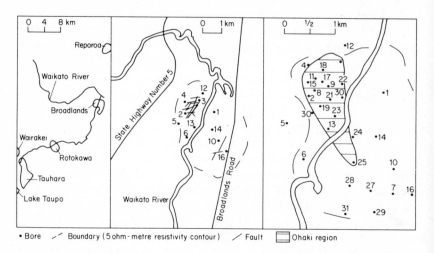

• Bore   ⌒ Boundary (5 ohm-metre resistivity contour)   ╱ Fault   ▭ Ohaki region

FIG. 7.8.   Broadlands geothermal field. Left, its general geographical setting; centre, some faults within it; right, the Ohaki region. Numbers are well numbers.

geothermal fluid at depth is believed to be neutral-pH chloride–bicarbonate, low-salinity water. The surficial escape of hot water is governed by recently active faults and proximity to the intersection of these is important in the siting of drill holes. A prominent interface at depths of 50–400 m was determined by a refraction seismic survey, an interface probably corresponding to the upper surface of buried rhyolite domes which may be related to the pattern of recent faults at the surface. The depth to basement graywacke was not determined, however, because of high seismic attenuation in the hot subsurface.[3] Owing to masking effects produced as a result of lateral variations of density in the volcanic sequence, the basement relief beneath the field could not be determined by means of a gravity survey. Electric and other surveys have indicated the actual boundaries.

A number of wells have been drilled into the field in the region of the resistivity low and mostly give temperatures even greater than 270 °C, but, as a consequence of low permeability, some only give low yields.

At Wairakei, there is good pressure communication between bores and the bore pressures have actually fallen in unison following a typical single field pressure, but at Broadlands, there is much greater variation between bores. Discharge enthalpy is more variable and so too is the pressure-drop factor which may vary widely from bore to bore. When exploitation ceased, some bores have shown pressure drops of up to 14 bars (200 psi), whereas, when a comparable amount of fluid has been withdrawn at Wairakei, there was very little drop. These drops refer to stabilised closed-in pressures. Occasionally, some bores at Broadlands do move together, namely those in the Ohaki area. Figure 7.8 illustrates where this is in relation to the entire field. Even here, however, they are not as closely coupled as are the Wairakei bores. Outside this area, the other bores are isolated from each other and also from the Ohaki bores. Couplings between the Ohaki bores are not perfect and it seems to take about a year for pressure to propagate across the area. This area is also isolated vertically as well as in a horizontal sense. One deep bore, BR15, is cased down to 1700 m and shows almost no effect of exploitation, thus indicating that the pressure drops of the other bores with typical working levels of 600 m or so have not attained this depth.

Around the 300-m depth, there is a 'cold river', bores 13, 17, 18 and some others showing this. The size of the cold anomaly decreases northwards and the interpretation is that it represents a natural cross-flow of cold water estimated at a few litres per second. An interesting aspect is that the exploitation does not appear to have affected this feature. In fact, the Ohaki area includes all the good producing bores and M. A. Grant presented a model of the response of this subfield to exploitation and its recovery.[4] Total bore discharges show gas fractions of from 0·5 % to 10 % by weight and there is a degree of correlation between gas fraction and bore enthalpy, higher enthalpy implying higher gas fraction. Maximal temperatures range up to over 300 °C. Actually, from the initial boring activities (1966–1971), it was expected that poor permeability conditions would exist except to the west of the river and, of course, almost all of the area is, in fact, west of it. To 1977, Broadlands was drilled up to a proven 150 MW and the 1975 Power Planning Report recommended the installation of a plant of this capacity for commissioning in 1981. Broadlands constitutes the third geothermal field substantially to be developed in New Zealand and Grant concluded that the small percentage of CO_2 occurring has had a major role in determining its future as an important geothermal source.[4]

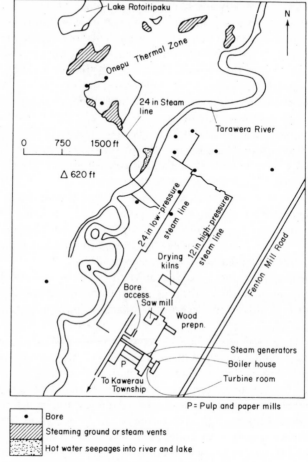

FIG. 7.9.   Kawerau geothermal steam field.

7.3.   OTHER GEOTHERMAL FIELDS IN NEW ZEALAND

At least 19 other geothermal fields exist, but not all are suitable for development. They include Orakeikorako, Reporoa, Rotokawa, Tauhara, Te Kopia, Waiotapu, Waikiti and Tikitere in the thermal belt and Ngawha Springs in the extreme north. At Kawerau and Rotorua, direct utilisation is going on. In the former case, exploration began in 1952 and several wells have been drilled, steam from these being utilised for heat-exchanging with boiler-quality water in order to generate high-quality steam for mill processes. At Rotorua, more than a thousand hot-water wells supply thermal energy to individual homes as well as to schools, hospitals, industry and hotels, one hotel with 100 rooms being geothermally air-conditioned. The wells are mostly shallow (150 m or so) and the temperatures range between 130 and 175 °C. They produce a mixture of steam, hot water and non-condensable gases. One distinction from Iceland (v. also Chapter 8) is that there, space heating is operated by the municipality, whereas in New ✕ Zealand this is not the case (although subject to government regulations).[5] The above examples illustrate the fact that geothermal heat may be used more efficiently in heating than in electrical power generation.

The steam-field layout is illustrated in Fig. 7.9.

REFERENCES

1. Bolton, R. S. (1977). Geothermal energy in New Zealand. In: Geothermal energy technology. Section 7 in: *Energy Technology Handbook*, ed. Douglas M. Considine. McGraw-Hill Book Company, New York, pp. 14–38.
2. Smith, J. H. (1970). Geothermal development in New Zealand. *Geothermics, Sp. Issue*, **2**(2), 424–36.
3. Hochstein, M. P. and Hunt, T. M. (1970). Seismic, gravity and magnetic studies, Broadlands geothermal field, New Zealand. *Geothermics, Sp. Issue*, **2**(2), 333–46.
4. Grant, M. A. (1977). Broadlands—a gas-dominated geothermal field. *Geothermics*, **6**, 9–29.
5. Koenig, J. B. (1973). World wide status of geothermal resources development. In: *Geothermal Energy*, ed. Paul Kruger and Carel Otte. Stanford University Press, Stanford, Ca., pp. 15–58.

CHAPTER 8

Space and Process Heating

The generation of electrical power is one of the major applications of geothermal energy, but, as noted in the preceding chapter (section 7.3) there are others, notably space and process heating. These and others are listed in Table 8.1 which shows them *vis-à-vis* the temperature range of most geothermal water and steam.[1] From this table, it is apparent that space heating and greenhouses fall into the relatively low-temperature category while process heating belongs to the upper one.

Where natural steam is available, usually at pressures ranging from 5 atm to 20 atm, exploitation near the source is desirable because, owing to the high volume of steam, pipelines are expensive to construct. Additionally, there is a pressure loss after long-distance transportation and this may well decrease its utility as a heating agent. To keep piping short, it is necessary, wherever possible, to use the steam nearby, if feasible actually within the geothermal area. Sometimes, the economics of process-heating applications are such that it may well be more practical to transport raw materials *to* the steam source rather than attempt to convey the steam to them.

Steam pipelines may be made of low-carbon steel and incorporate means of expansion. They are insulated with glass wool or other suitable material covered by an aluminium jacket.

Water is separated where the steam is wet and this may be done by means of special cyclones located at the borehole. Each phase is piped separately. Nevertheless, two-phase flow in pipelines is possible.

If long-distance transport is necessary, the water only is dispatched. In Iceland, distances of 16 km have been involved for decades.[1] It is very interesting to note that it has been found that the transportation of heat as water may be considerably less expensive than the transmission of electrical power. This is even more true where high-energy loads are concerned. The

130

TABLE 8.1

APPLICATIONS OF GEOTHERMAL ENERGY WITH REFERENCE TO TEMPERATURE RANGE
OF MOST GEOTHERMAL WATER AND STEAM

Temperature (°C)	Application
	Process heating
180	Evaporation of highly concentrated solutions
	Refrigeration by ammonia absorption
	Digestion in paper pulp, kraft
170	Heavy water through hydrogen sulphide process
	Drying of diatomaceous earth
160	Drying of timber
	Drying of fish meal
150	Alumina through Bayer process
140	High-speed drying of farm products
	Food tinning
130	Evaporation in sugar refining
	Evaporation of salts by evaporation and crystallisation
120	Fresh water by distillation
	Most multiple-effect evaporations, concentration of saline solution
	Refrigeration by medium temperatures
110	Drying and curing of light-aggregate cement slabs
100	Drying of organic materials, seaweeds, grass, vegetables, etc.
	Washing and drying of wool
90	Drying of stock fish
	Space heating
80	Greenhouse by space heating
70	Refrigeration by low temperature
60	Animal husbandry
	Greenhouse by combined space and hotbed heating
50	Mushroom growing
	Therapeutic baths
40	Soil warming
30	Swimming pools, fermentations
	Warm water for all-year mining in cold climates
	Deicing
20	Hatching of fish
	Fish farming

Notes at right (braces spanning 140–180): Temperature range of conventional electrical power production

Left margin labels: SATURATED STEAM (180–90); WATER (80–20)

temperature of water in pipelines is frequently below 100 °C, but can be as high as 125 °C. G. Bödvarsson and J. Zoëga suggested that, under suitable conditions, the optimal temperature from a cost point of view may be between 160 and 180 °C.[2] Hot-water supply pipelines are normally made of ordinary steel, insulated, and major lines are protected by concrete conduits. Multistage centrifugal pumps maintain the flow rate of the water.

TABLE 8.2

DATA APROPOS CROSS-COUNTRY HOT-WATER SUPPLY PIPELINES FOR DISTRICT HEATING
SYSTEMS IN ICELAND

Characteristics	Reykir–Reykjavik		Hveravellir–Husavik supply conduit
	Main supply conduit	Older supply conduit	
Type of water involved	Geothermal	Geothermal	Geothermal
Length, km	13	16	19
Temperature of water, °C	86	86	98
Diameter of pipeline, mm	700	Two 350	250
Capacity, tons/h	5 000	1 000	200
Pumping power, kW	1 800	700	Self-flowing
Material of pipe construction	Steel	Steel	Cement asbestos
Type of insulation	Rock wool	Turf	Earth
Temperature loss, °C	1·5	5	18
Protection method	Concrete conduit	Concrete conduit	Earth cover

Table 8.2 gives details regarding some cross-country hot-water supply pipelines for district heating systems in Iceland taken from Baldur Lindal.[1]

8.1.   SPACE HEATING

Iceland is a country which uses geothermal energy for space heating to a very great extent. Almost half the population have houses heated from this source and plans are in hand to extend the coverage to 70 % of the people within the next few years. The temperatures of the geothermal fluids which are involved range from 60 °C to as much as 150 °C and this is interesting because many parts of the world have geothermal reservoirs at economically acceptable depths containing fluids within this temperature range. This means that such heating applications are possible in many other places (*v.* Fig. 8.1). Obviously, although heating by such means may be applied on a small scale, even to a single home in a rural area, a wider distribution—to a city perhaps or even an entire region—is a commoner approach. The utilisation of geothermal energy for space heating is excellent from an environmentalist's point of view because there is no smoke emission and warm effluents are widely distributed to the sewage

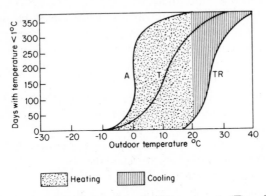

FIG. 8.1.   Thermal characteristics of arctic (A), temperate (T) and tropical (TR) climates. From S. Einarsson (1973), Geothermal district heating, in *Geothermal Energy*, UNESCO, New York, p. 129.

system. The cost of energy pleases the economists also because it is very low indeed compared with that of fossil fuels. The depreciation time for equipment also is very favourable and figures of up to 30 years are normally considered suitable.

8.1.1. Distribution

Where hot water is the relevant medium, single-pipe systems are usually used and water later discharged into the sewage system. Preferably, the temperature of the distribution water is in the 80–90 °C range and thereafter it cools to about 40 °C upon use. Supply mains to the distribution system will normally discharge into storage tanks which assist in coping with diurnal fluctuations in hot-water load. In order to maintain sufficient pressure in the distribution system, booster pumping is usually necessary. In towns, the distribution network is installed underground in the streets. Those street mains which have diameters exceeding 3 in can be placed in concrete channels and are insulated by some appropriate material such as rock wool (glass wool) or aerated concrete. The channels are themselves embedded in gravel together with concrete drainpipes. The minimal inclination of these channels is kept to 0·5%. At street junctions, the channels may meet in concrete chambers where valves, fastening bolts and expansion joints are placed. Appropriate ventilation facilities are provided. Drainage is effected from the bottom and where this is not feasible, a pump pit is utilised. Smaller street mains and house connections therefrom may be insulated with polyurethane foam and protected by a jacket of high-density polyethylene which is water-tight.

Obviously the heating system of any district must be related to its climate. The most significant element of this is probably the daily temperature variation to be expected, but the annual range of temperature must also be considered. Overcapacity in the heating system must exist sometimes, naturally, because this must be adequate for the coldest period of the year. It has been found that, as a general rule, the cost of geothermal energy for space heating is closely proportional to the maximal required capacity. Methods are employed, therefore, to increase the annual load factor. For instance, this may be accomplished by designing the system for an outside temperature rather above that of the coldest day of the year. This involves boosting from other sources for a very short period annually. Another approach is to include within the system a fossil-fuel booster (this can raise the temperature of the water during very cold spells). Alternatively, the system may include a local geothermal subterranean reservoir where deep well pumps are installed in the drill holes. This set-up can yield increased production over these short periods by pumping at a draw-down of the water level.

8.1.2. Heating Homes

Central-heating systems are utilised and the hot water is admitted directly to these and discharged to sewage after employment. Inferential water meters with a magnetic coupling between the flow sensor and register mechanism are often utilised. The maximal flow of hot water is also controlled by a sealed, maximal-flow regulator. Occasionally, only minimal-flow regulators are used. If the hot water cannot be directly supplied because of high mineral contents which could cause a lot of scaling, heat exchangers may be used between the hot water and water which circulates in the central-heating system. Public buildings and industrial installations are also geothermally heated.

8.1.3. Agriculture

Greenhouse heating is the obvious application and this is effected in a

TABLE 8.3

EXISTING AND PLANNED APPLICATIONS OF GEOTHERMAL ENERGY FOR PROCESS USE

Product	Country	Applications	Form of geothermal energy
Pulp and paper	New Zealand	Evaporating, digesting, drying	Primary and secondary steam
Timber drying and seasoning	New Zealand, Iceland	Drying, seasoning	Steam, hot water
Diatomite processing	Iceland	Drying, heating, deicing	Steam
Seaweed drying	Iceland	Drying	Hot water
Wool washing	Iceland, USSR	Heating and drying	Steam
Curing and drying of building material	Iceland	Heating and drying	Steam, hot water
Stock fish drying	Iceland	Drying	Hot water
Salt recovery from sea water	Japan	Evaporation	Steam
Boric acid recovery	Italy	Evaporation	Steam
Brewing and distillation	Japan	Heating and evaporation	Steam
Planned:			
Hay drying	Iceland	Drying	Hot water
Salts from geothermal brine	Iceland	Evaporation	Steam

TABLE 8.4
PROCESS DESIGN FEATURES FOR GEOTHERMAL STEAM AND WATER

Operation	Geothermal steam		Geothermal water	
	Type	Examples	Type	Examples
Drying	Indirect heating	Steam tube driers, drum driers	Indirect heating	Multideck conveyor
Evaporation	Primary heat exchangers accessible	Forced circulation evaporators	Counter-current heaters	Preheaters
Distillation	Steam distillation	General equipment		
Refrigeration	Freezing	Ammonia absorption	Comfort cooling	Lithium and bromium absorption
Deicing	—	—	Direct application, indirect heating	Dredging aid, pavement deicing

number of countries. The effluent from space-heating systems may very well be used in small greenhouses and these produce crops of flowers, vegetables and seedlings.

8.2. PROCESS HEATING

Apropos process heating, there are three basic questions.[1]

1. What are the products which may use heat from geothermal fluids?
2. What are the advantages of using geothermal heat for these products as against competitive energy?
3. If there is a disadvantage as regards site location, can this be offset by the lower-cost energy?

Lindal opines that, because our present technology is largely tailored to using fossil fuels, no conclusive answers may be sought from current engineering and economic practice.[1] He gave valuable examples of existing and planned applications of geothermal energy for process use and these are appended in Table 8.3.

Lindal also summarised process design features for geothermal steam and water and these data are given in Table 8.4.[1]

Clearly, geothermal energy is a very versatile source of energy and it is, of course, possible to integrate process heating, space heating and electrical power production into a single overall system. This has been done, in fact, already. When the main objective of utilisation of geothermal resources is space heating, secondary electrical power generation is sometimes feasible. Other secondary applications include soil warming, heating swimming pools, greenhouses, etc. When process heating is the main objective and depending upon the geothermal source, some generation of required electrical power may well prove to be possible and there are usually plenty of chances to develop numerous heating applications.

REFERENCES

1. Lindal, Baldur (1977). Geothermal energy for space and process heating. In: Geothermal energy technology. Section 7 in: *Energy Technology Handbook*, ed. Douglas M. Considine. McGraw-Hill Book Company, New York, pp. 43–58.
2. Bödvarsson, G. and Zoëga, J. (1964). Production and distribution of natural heat for domestic and industrial heating in Iceland. *Proc. UN Conf. New Sources of Energy*, **3**, 452.

CHAPTER 9

Other Geothermal Regions

9.1. USA

The Geysers have been discussed already in Chapter 6 and the 7-mile zone in which they occur represents a vast geothermal resource exploited by an installation which, to date, is the largest geothermal electrical power plant in the world (a total capacity of 1000 MW is envisaged).

However, other promising areas exist in the USA and investigations have been accelerated by the passage of two important pieces of legislation by the US Congress. These were

1. The Geothermal Steam Act 1970, which considerably enlarged the development of geothermal resources in the western states, and
2. The Geothermal Energy Research Development and Demonstration Act 1974, which aimed at facilitating an assessment of the nature and state as well as the dimensions of the country's geothermal resource base.

Known geothermal resources areas on public domain in the western USA have been estimated to be almost 1·5 million acres (approximately $6 \times 10^9 \, m^2$). The western states involved are Alaska, Washington, Oregon and California along the Pacific coast, constituting, in fact, part of the so-called Pacific Girdle of Fire, a tectonically active, plate-marginal belt marked by volcanism. Hot-spring and fumarolic phenomena are hardly surprising in these states and also in Hawaii, therefore. However, they also occur in Idaho, Montana, Wyoming, Utah, Nevada, Colorado, Arizona and New Mexico.

At least 100 hyperthermal regions occur in these 13 states, i.e. regions in which the surficial temperatures are high and suggest a boiling fluid below

138

the surface. Interestingly, the Aleutian Islands of Alaska have as much hyperthermal activity as does Hawaii.

Of the geothermal resources of the USA, 90 % are believed to lie within these western states and to the 100 regions mentioned above may be added at least 1000 geothermally significant localities characterised by hot springs and fumaroles. The Pacific Gas and Electric Company put the first industrial unit on line at The Geysers in 1960 and this was the earliest that electrical power from geothermal sources was available in the United States. Areas under active exploration as to feasibility include Casa Diablo, Salton Sea, California; Beowawe, Brady's Hot Springs, Nevada; and Yellowstone National Park, Wyoming.[1] These have base temperatures of between 180 and 200 °C and comprise hot-fluid or steam resources.

In Marysville in Montana, investigations are proceeding into hot rock geothermal resource development and here, at 2000 m or so depth, the temperatures attain 425 °C or more.

9.1.1. Yellowstone National Park
The Yellowstone National Park may well be the largest and indeed the hottest geothermal region in the USA, the entire district extending for 40 miles (Fig. 9.1). Several subfields exist. In the north, the Norris Geyser

FIG. 9.1. Thermal springs of the USA.

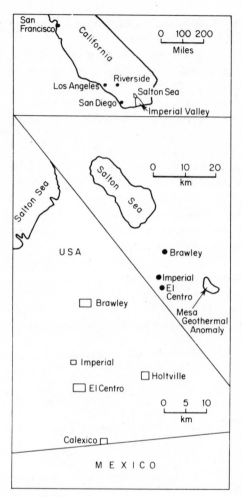

FIG. 9.2. The geographical setting of the Salton Sea and Imperial Valley in southern California, USA.

Basin is found and contains at least a dozen geysers as well as geothermal pools. North again occur the Mammoth Hot Springs and in the south is to be found the famous Old Faithful Geyser.

Shallow wells have been drilled and, at 300 m or so, they attain temperatures of 280 °C. One well has yielded dry steam and others have produced hot water which has flashed to steam. Ash flow formations, together with basalt flows, obsidian, glacial debris and diatomaceous sediments exist and suggest a volcanic episode which extended into the Holocene.

9.1.2. Imperial Valley, California (v. also Chapter 10)

The Imperial Valley lies in the Gulf of California and just over the East Pacific Rise and the North American continent (Fig. 9.2). Almost 20 000 ft of sediments occur and convey a heat flow 10 times average, indicative of extremely high temperatures deep in the basin so that, in the water, dissolved solid content has attained 260 000 mg/litre. The ground water actually recoverable is estimated to be 1.1×10^9 acre-feet with a temperature of 150 °C or more and dissolved solid content under 35 000 mg/litre. Clearly, this last factor is important because, where dissolved solid content is too high, the cost of the pretreatment water becomes prohibitive. Almost two-thirds of the recoverable ground water originates in the Colorado River, the rest coming from local sources. Exploration activities include feasibility studies on the development of a dual exploitation system both providing desalination facilities and also generating electrical power.

9.1.3. Salton Sea, California (v. also Chapter 10)

Between 1927 and 1965, 15 wells were drilled in the Salton Sea area, both for oil and geothermal purposes. The surficial phenomena include boiling mud pots and seeps of CO_2, and enthalpy is given as about 450 Btu/lb. The brine contains around a quarter of a million parts per million of chlorides, these being of Na, K and Ca. There is about 20 % flashover to steam. A 3-MW pilot generator has been constructed.

9.1.4. Casa Diablo, Mammoth, California

Eleven wells were drilled at Casa Diablo between 1959 and 1969, entirely for geothermal purposes. Fumaroles and hot springs occur on the surface. The depths of drilling were to a thousand feet or so, compared with 8100 ft at the Salton Sea. Hot water is supplied with 5 % steam flashover. The Magma Power Company and its associates were involved in these activities.

9.1.5. Beowawe, Nevada

Eleven wells were drilled at Beowawe for geothermal purposes between 1959 and 1965 by the Magma Power Company. Hot springs, geysers and fumaroles constitute ground manifestations. Depths of wells extended to around 2000 ft. The hot water has 5–10 % steam flashover.

9.1.6. Brady's Hot Springs, Nevada
Again, 11 wells were drilled to maximal depths of over 5000 ft by the
Magma Power Company and associates during the period 1959–1965 for
geothermal purposes. Boiling springs and seepages of steam occur
surficially. The hot water has 5% steam flashover.

9.1.7. Steamboat Springs, Nevada
From 1920 to 1962, 36 wells were drilled at Steamboat Springs to maximal
depth of 1830 ft in connection with geothermal investigations and irrigation
water by the USGS and the Magma Power Company and associates.
Surface manifestations include boiling springs, geysers and fumaroles.

9.1.8. Clear Lake (Sulphur Bath Mine), California
Hot springs and hot mine waters occur in the Clear Lake area and four
geothermal exploration wells were drilled by the Magma Power Company
and its associates as well as Earth Energy Inc. from 1961 to 1964, the
deepest penetrating to around 5000 ft.

9.1.9. Other Californian Prospects
These include Surprise Valley (Lake City), Calistoga, Wilbur Springs,
Amedee, Kelley's Hot Spring, Randsburg and Fort Bragg.

9.1.10. Other Nevada Prospects
Other Nevada prospects include Darrough, Monte Neva, Genoa, Pyramid
Lake and Crescent Valley.

9.1.11. Oregon
An example is Lakeview where, in 1961, two wells were drilled to maximal
depth of about 650 ft. Hot springs occur surficially. Geysers are also present
and occasionally erupt to 50–60 ft. Hot water is utilised in greenhouses.

9.1.12. Hawaii
An example is Puna (Kalapana, Pahoa) where seepages of steam
associated with the 1959–1960 eruptions of Kilauea occur in the eastern rift
zone. Five wells were drilled in 1961 to a maximal depth of 692 ft.

9.1.13. Salt Domes
If salt domes can be sufficiently mobilised to flow, low-density rocks will
move upwards through other rocks and form piercement structures
(diapirs), migrating in circular blobs anything from 1 km to 10 km across.

As they get near the surface, these blobs can exert enough force to buoy the overlying rocks into domes and break them up into distinctive patterns of faults. Salt often penetrates the surface and it may there become dissolved to form a shallow basin ringed with ridges of eroded rock. This is especially likely in humid regions. Faults associated with diapirs typically form along vertical radial lines or about concentric cones above the summit of the ascending blob. This is, of course, what might be expected from a tensional fracture above a point.

Salt domes are usually proximate to deposits of petroleum and gas and may be mined for potash as well as salt. Their most interesting characteristic from the present point of view is their possible importance as sources of geothermal heat.[2] They constitute a geological heat anomaly and this may be seen from Table 9.1.[1]

A typical salt dome (Fig. 9.3) occurs in the Lovann Formation of the Gulf Coast. In this, at a depth of 14 552 ft, a temperature of 240 °C has been recorded. At 25 000 ft, it is estimated that the temperature would reach 300 °C. Of course, such depths may well prove to be totally uneconomic and are far greater than any utilised to date in the extraction of geothermal energy. The heat flow is estimated to be anything between six and ten times greater than that of the surrounding area. Heat from such a salt dome might well be made available. It is a matter of advanced technology.

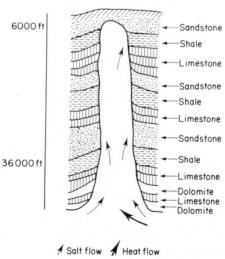

$\nearrow$ Salt flow $\nwarrow$ Heat flow

FIG. 9.3.   Subterranean cross-section of a salt dome (Lovann, Gulf Coast, USA).

TABLE 9.1

COMPARISON OF SOME ROCK TYPES WITH ROCK SALT

Rock type	Thermal conductivity ($\times 10^{-3} cal/cm\,s\,°C$)
Sedimentary rocks	1–8
Igneous rocks	3·1–9·8
Metamorphic rocks	5·2–8·4
Rock salt	17

9.2. MEXICO

The Federal Electricity Commission has effected quite an intensive geological, geochemical and geophysical survey of the region around Cerro Prieto. The American side of this is under investigation also and it constitutes part of a geothermal field related to the San Andreas fault system. Reservoir temperatures are in excess of 300 °C and the depth of the average well is 4000 ft. To 1976, 22 were drilled and from them are derived a high-temperature, hot-water system and therefrom a flash-steam powered electrical generating plant with a capacity of at least 37·5 MW.

Expansion plans are in hand.

9.3. ICELAND

A 3000-kW geothermal power station exists in the northern part of Iceland at Namafjall and the utilisation of this together with other geothermal energy is equivalent to combusting 1170 MW of electrical power. Expansion programmes are being conducted and so is additional exploration.

9.4. USSR

There is a great commitment to geothermal energy in the USSR and work is proceeding on

1. the study of deep thermal processes,
2. the regional distribution and conditions of formation of geothermal fields,

3. the development of technology,
4. the practical uses of deep heat.

Some volcanic areas in the Soviet Union have 100 °C waters at the surface and, at 100–400-m depth, some parts of Kamchatka and the Kurile Islands have temperatures up to 200 °C. Actually, it is believed that over half of the total territory of this vast country is underlaid by economically exploitable thermal waters. The geothermal water reserve is estimated at $7.9 \times 10^6 \, m^3/day$ at a depth of 1000–3500 m and with temperatures between 50 and 130 °C, almost three-quarters of this being 1500 m from the surface.

9.5. ELSEWHERE

Other areas include Ethiopia, Indonesia, Western Anatolia, Iran, Nicaragua, Taiwan, Colombia, Guatemala, Costa Rica, Uganda, Tanzania, India, Czechoslovakia, Yugoslavia, Hungary and Poland.

Development in these countries is very variable and will accelerate as a result of UN efforts and the earmarking of $US185 million as an initial investment by the US government to 1980.

9.5.1. Great Britain
Since completion of this book, details have come to hand regarding important work at Harwell's Energy Technology Support Unit which has led to the conclusion that suitable geology exists to support thermal waters in hitherto unsuspected parts of this country such as Southampton, Liverpool, Cheshire, Lincolnshire, Yorkshire and even Northern Ireland. Dr John Garnish of the unit believes that such resources could well be economic compared with normal fossil fuels. The Department of Energy is drawing up plans for a National Geothermal Board and has approved a £1·5 million scheme aimed at drilling for such waters up to depths of several miles.[3]

9.5.2. Tibet
Although Lamaism regarded geothermal lakes and springs as creations of the Buddha and hence untouchable, peasants and herdsmen in Tibet have utilised these geothermal sources for centuries, according to Liao Zhijie *et al.*[4]

The total natural heat flow of the region is estimated to be 680 000 kcal/s, a potentially enormous energy potential making this country the primary region of this type in China, into which it is now forcibly incorporated. Geysers have been located in three places at Bibilung in Namling county, central Tibet, and a hot, briny spring exists in Yanjing (salt wells) to the south of Markam county, eastern Tibet. Here, wells have depths of from one to several metres and contain 3% hot brine concentrations with temperatures ranging from 26 °C to 41·4 °C, compared with 14 °C in the adjacent River Lancang.

The town of Cona is built on a northern slope of the eastern part of the Himalayas and hot spring water is channelled into a bathing pool in every home. It is also used for heating so that, at an altitude of 4600 m above mean sea level, the community enjoys indoor temperatures of 18 °C even when the cold outdoors is well below freezing. Hot Spring Valley (Wenquangou) is in Zogang county, eastern Tibet, and here the lightly mineralised geothermal waters range in temperature from 32 °C to 38 °C. The first geothermally heated greenhouse and swimming pool in the country have been built in Xaitongmoin county, southern Tibet, 4000 m above mean sea level. Finally, 90 km northwest of the capital, Lhasa, is a fair geothermal potential on the Yangbajain Basin, 7 km² in extent. To the east, there is a hot lake from which a steam column rises about 100 m high. The natural heat flow here is thought to be 110 000 kcal/s and temperatures as high as 165 °C are reported at 60-m depth. In 1977, a 1-MW experimental power station was inaugurated at Yangbajain and this utilises the steam from a geothermal well to drive its turbines.

REFERENCES

1. Cheremisinoff, Paul N. and Morresi, Angelo C. (1976). *Geothermal Energy Technical Assessment*. Technomic Publishing Co. Inc., Westport, Conn., 164 pp.
2. Jacoby, C. H. (1974). Salt domes as a source of geothermal energy. *Min. Engineering*, **26**(5), 34–9.
3. Elliot, Harvey (1979). Home affairs. *Daily Mail*, March 31 issue, p. 13.
4. Zhijie, Liao, Zhifei, Zhang, Maozheng, You and Mingtao, Zhang (1979). Heat beneath Tibet. *The Geographical Magazine*, May, pp. 560–6.

CHAPTER 10

Geothermal Resources and Water

Water is in short supply in many parts of the world as the writer noted during his UN work for the International Atomic Energy Agency in the Sahel countries of West Africa 10 years ago. Alleviating shortages or indeed averting them, if possible, is clearly very important and among the various solutions proposed are those of importation, reclamation and desalination. The difficulty is that all these require large quantities of power and unfortunately this is also likely to grow short in the near future. Also, increasing preoccupation with the dangers of pollution may compound the problem. This is one of the reasons why geothermal energy is of such great potential world-wide significance. It is practically the optimal clean source of power and also, where such energy resources exist, it is relatively cheap. Large amounts of water may be derived from these. Thus, after conveying thermal energy from reservoirs to the surface of the Earth, fresh water may result from steam-turbine condensates. Often, the liquid water emerging from geothermal wells is saline and here distillation is a ready method of making additional supplies of fresh water. Large-scale geothermal distillation is a technique which has been studied by a number of people. An alternative approach is to use the electricity generated by a geothermal power plant to effect desalination. This can be done using either sea water or brackish water, and membrane desalting is suggested as being more appropriate than distillation.[1]

The various geothermal systems have been alluded to in detail in Chapter 2 and from this it may be noted that the liquid-dominated reservoirs will be optimal for the obtaining of water supplies because they are possessed of large amounts of it at the beginning of the operation. There are many such reservoirs in various parts of the world. The vapour-dominated types such as The Geysers are obviously more limited as water sources. Liquid-dominated reservoirs offer a great potential for water *and* power production.

147

10.1.   CHARACTERISTICS

Those characteristics important to the present discussion comprise a suitable cap rock which is impervious and overlies strata of varying degrees of permeability. This last factor and differences in liquid density resulting from temperature variation will cause the geothermal liquid to circulate as V. E. Schrock and his associates demonstrated in the laboratory.[2] This is a very slow process, of course, and it transports hot liquid quite close to the surface where it is tapped by producing wells. From the flanks of the reservoir, cool water migrates in as a replacement fluid after its injection through suitable wells. This follows the natural convection currents away from the hot zone near the surface. The intention is that the injected water will enter the hot part of the reservoir at depth and later be withdrawn as hot fluid through the producing wells. This model entails the piping of the hot withdrawn geothermal fluid to the power and water plants; fresh water not locally needed would, like the electrical power generated, be exported. Disposal of reject brine could be effected by release to surface water or by evaporation or by returning it underground either through a deep well injection or by simple seepage.

An open system approach is one in which the geothermal fluid is separated into two streams, one of brine and the other of vapour, the latter being fed through a turbogenerator in order to produce electrical power. As noted earlier, the condensate from the turbine exhaust can be utilised as a source of fresh water. If more is required, the brine may also be employed through desalination. This open system is already in use in New Zealand. Disadvantages can arise if corrosional effects result from gases being released or from undesirable precipitation of minerals. Where such phenomena are believed likely to accompany the phase change of a particular geothermal brine, a closed system may be substituted. The closed system uses a pump to maintain enough pressure in the brine to suppress phase change as it comes to the surface, passes through a heat exchanger and returns to the reservoir. Such a pump is placed near the actual producing part of a well. In the heat exchanger, the brine cools and heats a secondary fluid which is then employed to drive a turbogenerator. No water can be *directly* produced by means of a closed system. Another name for this is the vapour-turbine cycle.

Naturally, complex plants are necessary in order efficiently to recover water from brines. Table 10.1 lists some examples demonstrating the effects of various system parameters and selections of processes.[1] These examples are based upon a production rate of 1 million lb/h of well fluid.

TABLE 10.1

APPROXIMATE CALCULATED WATER AND POWER PRODUCTION FOR SIX CASES UNDER
VARIOUS ASSUMED CONDITIONS

Production parameter	Case 1	Case 2	Case 3	Case 4	Case 5	Case 6
Reservoir temperature (°C)	315	315	315	315	205	150
Wellhead temperature (°C)	229·2	229·2	229·2	229·2	150	93·3
Wellhead pressure (psia)	400	400	400	400	67	11·5
Lowest temperature (°C)	100	71·1	71·1	71·1	71·1	71·1
Fresh water/well fluid ratio	0·79	0·83	0·79	0·87	0·43	0·26
Water production rate at 90% plant availability per million lb/h of fluid supplied						
Million gal/day	2·04	2·15	2·05	2·26	1·13	0·62
Thousand acre-feet/year	2·29	2·41	2·30	2·55	1·27	0·70
Power production rate (MW)	17·7	24·3	24·3	24·1	6·7	1·2
Percentage of Case 2						
Water rate	95	100	95	105	53	29
Power	73	100	100	99	28	5

Laird gave a simplified flow diagram and thermodynamic-process diagram for Case 1 and almost 80% of the well fluid is converted to fresh water.[1]

The brine returned to the reservoir becomes 4·7 times more concentrated than the well fluid and, as seen in the table above, the calculated power-production rate is 17·7 MW; the water-production rate being at 90% availability (operating 0·9 year every year) is some 2 million gal/day, i.e. 2290 acre-feet annually. The fresh-water product is actually above its normal boiling point and therefore must be further cooled. This could be accomplished either by evaporation or by mixing with cooler water. Of course, there will be a loss of geothermal energy involved.

Case 2 indicates the effects of temperature reduction relevant to rejection of thermal energy and Case 3 is a simplified version of Case 2. All three cases have the disadvantage that non-condensable gases pass through a good deal of the equipment and could cause corrosion. At high pressures, non-condensable gases can be removed by mixing them with some of the liquid being injected into the reservoir.

Case 4 is a variation of Case 2, more efficient in that it does remove the non-condensable gases. In Case 5, it was assumed that the reservoir's thermal content is equivalent to that of liquid water at 200 °C and also that the fluid enters the system at a wellhead temperature of 150 °C. Lower temperatures are involved in Case 6. Penalties are incurred for lower reservoir temperatures and also for using higher waste-heat rejection temperatures. Even a small re-employment of the heat of condensation of the vapours results in a definite improvement in water production.

10.2. MULTIPLE-EFFECT AND MULTISTAGE DISTILLATION PROCEDURES

Multiple-effect distillation is carried out in such a manner that each event occurs at a lower temperature than the preceding one and it is the oldest method of desalination, modern plants of this type employing as many as 25 effects which can produce anything up to 20 lb of water from every pound of steam condensed to heat the initial step of the process. The US Bureau of Reclamation in 1972 set up an interesting application consisting of a three-stage vertical tube evaporator operating at about 15–16 °C from the assumed wellhead temperature of 200 °C. The incoming mixture is assumed to comprise 20 % steam and the first effect is similar to that of Case 4 cited above, including the removal of non-condensable gases. Two additional effects are also similar. The brine from effect 3 is five times more concentrated than the wellhead mixture, i.e. four-fifths of the wellhead fluid is converted to fresh water. The steam from the third effect is transmitted through a turbogenerator in order to produce approximately 10 MW.[3]

The multistage flash process of distillation has also been proposed as suitable for desalination of geothermal brines and in this, hot brine is passed through a set of chambers, each of which is at a lower pressure than the one preceding it. In fact, one step of the process takes place at each stage. Some of the brine is flashed as it enters each chamber. Physically, the chambers are each several feet in length because, since evaporation is not instantaneous, time must be allowed for the vapour to separate as the brine flows along the bottom of each chamber. In each chamber are tubes inside which cooler brine flows towards the hot end of the set of chambers. Vapour produced condenses on these tubes and drains off into a trough underneath them. The distillate is also passed through the chambers in the same way as the hot brine and its vapours also condense on the tubing and again add to the heat available for the cool brine. Industrial multistage flash sea-water

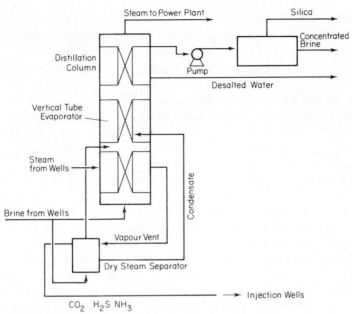

FIG. 10.1. Schematic of a desalination cycle.

distillation plants for desalination possess 30 or even more stages and here, spent thermal energy is rejected to the ocean.

The multistage flash distiller is perhaps the commonest technique for sea-water conversion and it has been estimated that about 90 % of all planetary desalination is effected using it.[1] Its advantages depend upon the fact that the hot brine channel can be so constructed that it will not get blocked even if solid materials such as silica precipitate in it; also, deposits adherent to the walls can be scraped off quite cheaply. Then too, the feed water does not need extra heating in geothermal work; with sea water, this is not the case, of course. Here, a heater must be placed between the discharge ends of the condenser tubes and the entrance to the series of flash chambers. The advantage here of the geothermal application is balanced by a disadvantage, however. In sea-water desalination, the cool brine inside the condenser tubes is incoming feed water in the process of being heated. In geothermal desalination, no substitute appears available for this excellent means of *cooling* the condenser (which otherwise would permit quite a high fraction of geothermal brine to be desalted). As a result, the multiple-effect approach is probably more suitable than the multistage flash one.

Figure 10.1 shows the typical layout for a desalination cycle. Currently, the cost of producing fresh water by means of large distillation plants is dropping and by 2000 A.D., should not exceed a dime or so per 1000 gal.

10.3. COST MANAGEMENT

Thermal energy is not storable, but must be converted to other forms of energy such as electrical or chemical which can be stored. Unfortunately, some of the energy obtained from a geothermal reservoir is not convertible and this must either be transferred into materials at lower temperatures or else radiated away, i.e. lost. Such waste heat enters the environment, sometimes without effect. In some cases, it may be beneficial and cause thermal enrichment. In other cases, it may have a deleterious effect and cause thermal pollution. Disposal is important, therefore, and the manner in which this is done is an important planning consideration in the use of geothermal resources.

In fact, the quantity of thermal energy transformable to chemical energy as a result of removing salt from saline solution by a perfect process is about 4 kWh per 1000 gal of water converted from sea water. This is small in relation to the geothermal energy which can be converted to electrical power or returned to the reservoir in reject brine or disposed of in some other way. The inference is that if water alone is being produced, i.e. if no attempt is made to return thermal energy to the reservoir, then almost *all* of the geothermal energy extracted constitutes waste heat!

If electricity is exported from a geothermal field area, desalination activities can be decentralised also and this can be an advantage to a region.

The main desalination processes operating by using electricity are vapour-compression distillation, reverse osmosis and electrodialysis. All are employed commercially and each is particularly suited to its own special circumstances.

1. Vapour compression

This is based upon the fact that the temperature of water vapour can be raised if a compressor is utilised to raise the pressure of the water vapour.

A temperature difference is thus made available and this can be used in order to promote the condensation of the compressed vapour on one side of a heat-transfer surface and thereby effect the evaporation of an equal quantity of water from brine on the other side of the said surface. The latter is then fed to the compressor. The actual temperature and pressure difference is kept small so as to minimise the power of the compressor and the staging principle can be incorporated into vapour compression plant designs.[4] The process can be made cheap and is, therefore, highly competitive. This is especially the case in lower capacity (up to 0·5 million gal/day) sea-water conversion activities. However, it is not suitable for feed waters such as geothermal brines which contain silica in appreciable amounts and also it is not competitive with reverse osmosis or electrodialysis in desalinating brackish waters.

2. Reverse osmosis

This is sometimes called hyperfiltration and it depends upon the phenomenon of osmosis, i.e. the flow of water through a semi-permeable membrane (a membrane allowing the passage of the solvent, but not of dissolved substances). Desalination depends upon most of the salts being held back by such membranes when a pressure difference higher than the osmotic pressure is maintained across them. The osmotic pressure increases with concentration difference across the membrane and therefore both the necessary energy and the cost of the water produced increase with the salinity of the feed water. Many successful variants of this are in service. The process is optimally applied to saline waters in the concentration range from 2000 ppm to 5000 ppm. Actually, within this range, the cost of desalination should be about 50% of that necessary if distillation is employed.[1]

3. Electrodialysis

Desalination by this method depends upon using two kinds of membrane which are selectively permeable to ions. One will pass cations better than the other (which passes anions preferentially). Alternating cation and anion membranes are arranged on the walls of parallel channels and saline water is flowed through these, an electric current being passed across. Obviously, the latter will carry ions of opposite sign through the membranes on opposite sides of each channel so that the concentration in every second channel is reduced, being increased in alternate channels. Geometrically, the channels must be made of sufficient length to ensure that the required reduction in concentration of the water produced is obtained. Theoretically, the electric current flowing across is proportional to the number of ions it transports, hence to the required salinity reduction. It is clear that the more concentrated the saline waters are, the more electrical power will be needed in order to achieve desalination. Electrodialysis is considered to be most useful for freshening brackish waters having concentrations up to 3000 ppm. This is a low range of concentration and hence costs are also low.

10.3.1. Budget

As regards water, it is important to note that, where large-scale operations are involved, it may be economical actually to import water for pressure maintenance and cooling. This has been proposed for the Imperial Valley in California, a geothermal area in a very arid region not particularly near to the coast. The exportation of as much as 2·5 million acre-feet annually of desalted water from the valley to the Colorado River was envisaged and the suggestion made that sea water should be imported in order to compensate, this being done from the Gulf of California about 100 miles or more away.[5] In smaller developments, such water might be locally available—one source might well be irrigation-drainage water. In fact, this has been proposed even in the Imperial Valley.[3] In this arid basin which is below sea level and devoid of a drainage outlet, the drainage sump is the Salton Sea which is maintained by agricultural drainage water. The water inflow is about 1·3 million acre-feet annually and this is balanced by the high evaporation from the surface of the saline lake (360 miles^2). Rex estimated that 150 000 acre-feet of water could be withdrawn every year and used to maintain pressure in geothermal reservoirs, and this would lower the level of the Salton Sea and also reduce its surface area until the evaporation reduction (12%) balances the withdrawal.[5] The US Bureau of Reclamation estimated that 125 000 acre-feet of water would have to be withdrawn from the Salton Sea in order to supply replacement and cooling water for their demonstration facility designed to produce c. 100 000 acre-feet annually of desalinated water and 420 MW of power. As regards solids, these have to be disposed of and, with salts, this can be done by injecting brine into subterranean void spaces produced by fresh-water production through desalination. Alternatively, such salts could be exported. The proposals for Imperial Valley would be highly beneficial because they would result in

1. stabilisation of both the level and the salinity of the Salton Sea;
2. reduction of irrigation-water salinity;
3. provision of cooling facilities for geothermal water and power plants;
4. prevention of surficial subsidence;
5. augmentation of the flow of the Colorado River.

It is thought that the benefits would be far in excess of the costs.

The Bureau of Reclamation study suggested that the unit cost of water delivered at a rate of 100 000 acre-feet annually would be between $US85 and $US130 per acre-foot for transportation distances less than 80 miles.[3]

Power from the associated 420-MW electrical power plant would cost 3 or 5 mills/kWh at the plant boundary (depending upon the source of funding) and the capital cost of the water plant would be some $US0·69 per daily gallon of capacity (comparable, in fact, to that for a multistage flash sea-water conversion plant of the same size as that currently operating at Rosarita Beach in Mexico). The electric plant cost would be approximately $US130 per kilowatt of capacity.

For the large-scale development in the Imperial Valley, the estimated unit cost for 2·5 million acre-feet annually of distilled water is between $US100 and $US150 per acre-foot delivered between 100 and 250 miles away (with sea water imported from a distance of 100 miles). Unit costs of 10 000 MW of electric power appear to be 3 or 5 mills/kWh and the estimated water cost at the plant site is about $US70 per acre-foot. For converted sea water from a plant of 10–50-million-gal/day capacity, this would be cheap.

Of course, all this is not meant to imply that the cost of any variety of desalination could compete successfully with large systems capable of supplying good water at a cost dependent only upon pumping costs from a proximate river. However, several desalination methods can be used to upgrade unusable (waste) waters such as saline ground water or sea water and these are already competitive with long-distance importation schemes. Undoubtedly, too, desalination costs will drop drastically before the end of this century.

REFERENCES

1. Laird, Alan D. K. (1973). Water from geothermal resources. In: *Geothermal Energy*, ed. Paul Kruger and Carel Otte. Stanford University Press, Stanford, Ca., pp. 177–96.
2. Schrock, V. E., Fernandez, R. T. and Kesavan, K. (1970). Heat transfer from cylinders embedded in a liquid-filled porous medium. Paper CT3.6, 4th International Heat Transfer Conference, Paris (Versailles). In: *Heat Transfer*, 1970. Elsevier Scientific Publishing Company, Amsterdam.
3. Bureau of Reclamation (1972). *Geothermal Resource Investigations, Imperial Valley, California*. US Department of the Interior.
4. Tleimat, B. W. (1969). Novel approach to desalination by vapor-compression distillation. *ASME*, Pub. 69-WA/PID-1. Annual Meeting of the American Society of Mechanical Engineers.
5. Rex, R. W. (1970). *Investigation of Geothermal Resources in the Imperial Valley and their Potential Value for Desalination of Water and Electricity Production*. Institute of Geophysics and Planetary Physics, University of California, Riverside, Ca.

CHAPTER 11

Geothermics in Italy

As has been noted earlier in this book, Italy is probably the most experienced country in the world with regard to geothermal energy and its development and this really began in 1904 when, in a Tuscan village, a small technical team were delighted to observe five electric lamps lit through the application of energy from steam emerging from a pit in the ground. Of course, this village was called Larderello and for many years it remained the sole place on Earth where the planet's internal heat was employed in power production. Actually, only a year later, the entire community was deriving its light from a 20-kW generator and continued drilling and exploitation enabled this figure to rise to 2500 kW by 1916 and 80 MW by 1940. In Italy, there is a marked lack of coal and there can be little doubt that this fact promoted rapid electrification of the railways. Here, the low cost of geothermal energy-generated electric power assured a market for that originating in Larderello. Naturally, research into this matter extended the area and so it is not surprising that by 1973 no less than 16 geothermal power stations with a total capacity of almost 390 MW had been installed in the region of Larderello and Monte Amiata (some 50 miles to the south).[1] It will be interesting to examine some recent work done on the Larderello geothermal field by C. Panichi and his associates.[2]

11.1.   ISOTOPIC GEOTHERMOMETRY AT LARDERELLO

Within the framework of an agreement between the Italian National Electric Energy Agency (ENEL) and the Italian National Research Council (CNR), a research investigation into the isotopic composition of CO_2 and CH_4 present in fumaroles, hot springs and geothermal fluids was effected over a decade by Panichi *et al.* with the objectives of determining whether the ^{13}C content of CO_2 showed any significant variation in relation to changes of the geothermal gradient and also of testing the validity of the CO_2-CH_4 isotopic geothermometer. Of course, CO_2 is by far the most abundant carbon compound occurring in the natural fluids which were analysed and the gas does show regional variations in isotopic composition, these being associated with thermal anomalies where higher $\delta^{13}C$ values are found.[3] It appears to be the case, therefore, that the carbon-13 content of natural carbon dioxide can be utilised as a secondary means of prospection in new geothermal regions.

Small quantities of methane and hydrogen always occur together with CO_2 in geothermal fluids and the assumption made is that isotopic equilibrium takes place between CO_2, CH_4, H_2 and water as a result of the following reaction:

$$CO_2 + 4H_2 = CH_4 + 2H_2O$$

The isotopic fractionation noted between the groups CO_2-CH_4, H_2O-H_2 and H_2-CH_4 have been employed in assessment of temperatures at depth in a number of geothermal regions, for instance by J. R. Hulston in 1977.[4]

11.1.1. The CO_2-CH_4 Isotope Geothermometer
For temperatures between 0 and 700 °C, Y. Bottinga has calculated the isotopic equilibrium constant for the system carbon dioxide–methane.[5] Hence, these data are available for what may be termed the geothermal range, i.e. from 100 °C to 400 °C. The relationship is expressible thus:

$$1000 \ln \alpha = -9 \cdot 01 + 15 \cdot 301 \times 10^3 T^{-1} + 2 \cdot 361 \times 10^6 T^{-2}$$

where $\alpha = (^{13}C/^{12}C)CO_2/(^{13}C/^{12}C)CH_4$ and T is the temperature given in K. It is only possible to apply this equation, however, in cases where it is believed that isotopic equilibrium exists between CO_2 and CH_4. There are grounds for thinking that to make such an assumption is invalid and these are outlined in a number of papers, for instance in that of B. D. Gunter and B. C. Musgrave published in 1971.[6] Panichi et al. indicate a number of relevant facts:

1. In all the geothermal fields looked into, this geothermometer gave temperatures usually 50–200 °C *higher* than those directly measured at the wellhead; it must be remembered, however, that the latter are often much *lower* than actual geothermal reservoir temperatures, *cf.* Ferrara and his associates.[7]

2. The CH_4–H_2 and H_2O–H_2 isotopic geothermometers, on the other hand, yield temperatures which are very close to those measured at the wellhead.[6]

3. CO_2 and CH_4 most probably coexist in isotopic equilibrium in the case where they derive from the same geochemical process.

4. Geothermal methane has a $\delta^{13}C$ usually ranging between -20 and $-30\%_0$ against PDB (Urey's original Belemnite standard), but methane originating at a lower temperature shows a much more negative value. This could mean that the isotopic composition of the methane reflects the high temperature of the formation. Alternatively, it could be that successive exchange with the CO_2 in the geothermal reservoir has occurred.

5. It is believed that the kinetics of the reaction

$$CO_2 + 4H_2 = CH_4 + 2H_2O$$

are extremely slow even at the high temperatures associated with geothermal fields.

6. At Larderello, it has been found that the isotherms given by the CO_2–CH_4 geothermometer parallel isotherms given by wellhead temperatures and this probably indicates that both sets move in a systematic manner.[2]

To reconcile the above facts is not at all easy and there are three possible interpretations:

1. In the absence of isotopic equilibrium, temperature evaluations from the carbon isotopic composition of CO_2 and CH_4 have no meaning, the bulk of the former originating differently from the bulk of the latter. However, this would imply that the parallelism between isotopic and measured temperatures alluded to above (6) is simply coincidental.

2. Isotopic equilibrium does exist, but some CO_2 and/or CH_4 of different origin are added to the system near the well bottom in a manner such that the equilibrium is upset and there is insufficient time for it to be re-established. This would not explain a number of facts, however, for instance why isotopic temperatures are significantly higher in the centre of the Larderello field where the likelihood of such additions taking place is negligible.

3. The isotopic composition of CO_2 and CH_4 reflects the temperature at depth where they are both formed together. When migration of

the geothermal fluid takes place and shallower regions are reached, successive cooling will take place, but this will not much affect the isotopic equilibrium established at depth because of the parallel drop in the isotopic exchange rate. This means that the isotopic equilibrium is frozen, as it were.

This could explain the difference between the isotopic temperature and the wellhead temperature and also the systematic shift of the first with respect to the second. Panichi et al. believe that this third hypothesis comes closest to the reality.[2] If they are correct, this has the important consequence that the temperatures given by the carbon isotope distribution between CO_2 and CH_4 do indeed reflect, at least to some extent, the temperatures which in fact exist in geothermal fields, particularly the temperature of formation of the two components. More investigation is necessary, especially as regards the experimentation required to confirm Bottinga's theoretical calculations on the fractionation factor between CO_2 and CH_4 in the geothermal temperature range.

11.1.2. CO_2–Water Vapour Isotope Geothermometer

It is a fact that the oxygen isotope exchange between CO_2 and liquid water is very rapid even at room temperature, hence an essential requirement for using such a geothermometer is to ensure that no significant re-equilibration takes place during the sampling process.

The equation of Panichi et al. for calculating isotopic equilibrium temperatures is

$$1000 \ln \alpha = -10{\cdot}55 + 9{\cdot}289 \times 10^3/T + 2{\cdot}651 \times 10^6/T^2$$

obtained by fitting the fractionation values between CO_2 and water vapour calculated by Bottinga in the temperature range 100–300 °C with $\alpha = (^{18}O/^{16}O)CO_2/(^{18}O/^{16}O)$ water vapour.[2,8] The results obtained show that, with a few exceptions, the isotopic equilibrium temperature of the geothermal wells is equal to or higher than the temperature measured at the wellhead. This may well show that the temperature given by oxygen isotopes corresponds to the actual temperature of the geothermal reservoir tapped (where CO_2 and water vapour are in isotopic equilibrium).

Further to pursue recent investigations in the area, it will be useful now to turn to Monte Amiata.

11.1.3. Seismic Noise Measurements

E. Del Pezzo and his associates have discussed seismic noise measurements with regard to the Monte Amiata region.[9]

It has been demonstrated in New Zealand and both at the Yellowstone National Park and Imperial Valley (USA) that high seismic noise amplitudes occur in the 1–20-Hz band. Prior to Del Pezzo's work, it was also shown in Italy at Lipari and additionally, by P. Cappello and his associates at Solfatara Crater, Phlegraean Fields.[10,11] This latter work is discussed below (section 11.2).

Monte Amiata is a vapour-dominated geothermal field and there is an excellent cap rock sealing an underlying permeable complex, adequate water supply and, of course, a marked thermal anomaly. A portable seismic station was utilised by Del Pezzo and his associates and some very interesting results obtained. First, a 3-Hz high noise level energy content is characteristic of the entire region. Peak frequencies in the 6–12-Hz frequency band occur only in some stations and may perhaps be ascribed to agricultural noise. Secondly, the energy decreases quite regularly moving away from exploited geothermal fields (at Bagnore and Piancastagnaio). For additional data on Monte Amiata, v. Chapters 2 (section 2.4) and 4 (section 4.1.2).

11.2. OTHER AREAS

It is important to realise that it is not only with the famous areas of Italy that investigation is concerned, but also with many others. For instance, a seismic noise survey was effected at Solfatara Crater, Phlegraean Fields by P. Cappello et al.[11] Here, weak thermal activity has been noted at the bottom of the crater and this is localised mainly in the southeastern zone which is characterised by fumaroles and boiling mud pools. A. Palumbo has demonstrated that maximal temperatures in the area range between 90 and 150 °C.[12] Cappello's survey was carried out in 1973 by measurements at 43 stations along some profiles, a closer grid being employed for the southeastern part of the Crater. Care was taken to distinguish variations in seismic noise caused by traffic and varying meteorological conditions. The frequency analysis of recorded seismic noise showed that the shape of the spectra changes considerably from station to station, but some characteristics were noted from which it was possible to group the data into three categories:

1.   Type 1 spectrum with higher than 20-Hz dominant frequencies;
2.   Type 2 spectrum with single or double peak between 8 and 16 Hz and dominant frequency sometimes in excess of 20 Hz;
3.   Type 3 spectrum intermediate between Types 1 and 2.

The most interesting fact is the lack of low frequencies in all spectra. Type 1 spectra originate in areas with maximal thermal activity and the single or double peaked Type 2 occur usually in the central part of the Crater (i.e. where *loose* materials attain their maximal thickness). Type 3 may well be determined by intermediate conditions and the mechanical properties of the surface layer. From these observations, it may be noted that fumaroles and mud pools are high-frequency and high-amplitude noise sources and also that seismic noise is strongly affected by the mechanical properties of the outcropping rocks. The workers concerned believed that perhaps a deeper source exists from which a lower-frequency seismic noise is emitted and this may be connected with a deeper and large-scale geothermal system.[11]

The Phlegraean Fields have also been subjected to a natural electric field survey by A. Rapolla.[13] This was part of a survey of three southern Italian geothermal areas and it will be convenient to discuss them all below.

11.3.   NATURAL ELECTRIC FIELD SURVEYS IN SOUTHERN ITALY

As well as the Phlegraean Fields, Lipari–Vulcano and Ischia were also involved. Measurements of the natural electric field vectors were carried out by recording the potential difference between two orthogonal electrode pairs with an electrode spacing of 20–50 m.

At the Phlegraean Fields (Fig. 11.1) west of Naples, measurements were carried out at 106 stations distributed almost regularly over the region. The intensity of the electric field turns out to be high, often exceeding 5 mV/m, but no obvious directional regularity is apparent. However, there was a correlation between areas of low resistivity and areas with highest values of electric field intensity. It must be added that a *local* decrease in field intensity was observed at places of surface thermal manifestations such as Solfatara. Curiously, there is no surface activity at Pianura, but high values were nevertheless recorded.

11.3.1.   Lipari and Vulcano
Lipari and Vulcano (Fig. 11.2) together with five other islands comprise the Aeolian group of the southern Tyrrhenian Sea and both belong to the same geological structural type, namely volcanic, rising from the floor of the sea to a total elevation of about 1500 m. The outcrops include pumices, tuffs, various pyroclastics, etc., and there is surficial thermal activity made up of

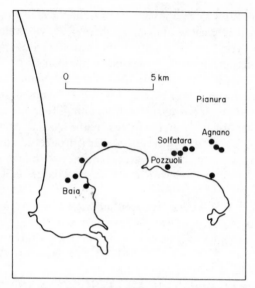

FIG. 11.1. The Phlegraean Fields.

fumaroles and hot springs, notably between Fossa and Vulcanello in the northern part of Vulcano. Thirty-eight randomly spaced stations were used in making measurements. The field intensity was found to be high everywhere (values usually of the order of several millivolts per metre). Again, no regularity was observed in the direction of the total electric field vectors. However, the highest field intensities were recorded around areas

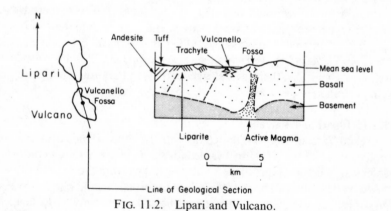

FIG. 11.2. Lipari and Vulcano.

with maximal thermal activity at the surface. A seismic noise survey noted high-frequency ground noise probably connected with the escape of gases at the surface along the vents of fumaroles and there is also a low-frequency ground noise which occurs over a much greater area (and probably represents a large geothermal system).

11.3.2. Ischia
Ischia is a volcanic island lying 30 km west of Naples and possessing outcrops of tuffs, loose pyroclastics and lava flows as well as several fumaroles and hot springs such as Casamicciola.

Measurements were made at 67 stations usually near the coastline. The same type of information was derived as in the above-mentioned islands, namely high intensity and no regularity in the field direction. However, here there is no correlation between areas of surface thermal activity and the occurrence of highest field intensity.

11.3.3. Some Inferences
The natural electric field survey results demonstrate that high electric field gradients occur at all of the surveyed sites, values as high as and sometimes even higher than 10 mV/m being found near the areas of maximal activity. Where averages over extended regions are taken, it has been noted that correlation between amplitude of electric gradient and thermal localities improves. In such a geothermal district, strong electric field gradients may be generated at the contact between media with significantly different electrical characteristics. Due to the high temperatures and salinity in geothermal systems, the electrical properties of their contained water will be markedly distinct from those found in water derived from outside such a system. Another interesting feature is that, as a consequence of alteration processes, even where the reservoir rock is originally homogeneous, the electrical characteristics develop considerable differences. Thus, patterns obtained are not dependent upon deep-buried sources, but rather appear to reflect discontinuities in the electrical properties of the shallow layers. In the Italian areas surveyed by Rapolla, surficial manifestations of geothermal activity are either quite localised or indeed totally lacking; nevertheless, considering the entire complex of shallow layers, such activity at the surface ought to be more extensive than in fact is the case.[13.] Consistent with the discussion is the pattern noted at Vulcano and at the Phlegraean Fields where the maximal values occur peripherally to the most active zones. At Ischia, a great irregularity in distribution of the electric field may possibly be the result of the presence of lava flows in the shallow subsurface of the

island, the flows acting as an additional disturbance to the geometry of the aquifers. It is extremely difficult to suggest a precise model capable of explaining in a quantitative manner the observed pattern of the electric field as a whole. This is because there are just too many parameters involved, some of them probably unknown. More data are urgently required. However, it seems clear that electric field vectors could develop into a suitable method for preliminary exploration of geothermal areas because they allow the individuation of shallow thermal manifestations and also permit the delimitation of their real horizontal extent.

11.4. ITALY—THE OVERALL PICTURE

Italian pioneering and efficiency in the geothermal resource field have been referred to already and it is certainly true that these have enabled the country to lead the world in research, exploration and exploitation even if the absolute amount of electrical power generated is less than that derived from The Geysers in California, USA. These activities in the Larderello/Monte Amiata areas are continuous and the former had 13 plants in 1976, these generating 365 MW in that year. At the same time, Monte Amiata (75 miles to the south) was generating 25 MW. In this complex, over 500 wells have been drilled, actually to an average depth of 500 m, and they yield an average of 23 000 kg/h at Larderello and 36 000 kg/h at Monte Amiata. This might be anticipated because the latter has been less time under exploitation and both mass flow and pressure decline with time. In fact, it is believed by some geologists that the declining yields and pressure at Larderello indicate that the field has already been tapped to capacity. In spite of this, research and exploration are continuing and new technological approaches are being applied, for instance infrared imagery is being utilised in order to determine whether any geothermal sources have been overlooked inadvertently. As mentioned above, there are many other areas in Italy which may prove to be useful. Here, for example at Monte Volsini, Monte Sabatini and Colli Albani, investigations are proceeding (v. Fig. 4.1). Northern Italy has been ruled out, but more southerly parts, notably around Mount Vesuvius, are interesting and here Compania Ovest is involved. The fluids are thought to be hot (naturally!) but not *dry* steam generators and so, in the event that they are to be commercially exploited, some new method of geothermal use and electricity production is necessary. Greater efficiency in turbines is always desirable. Early ones were of the non-condensable type which consumed twice as much steam per kilowatt-hour as do condensing steam ones (20 kg/kWh as against 10 kg/kWh). Obviously, the more effective use of the steam will greatly increase electricity output from the same quantity of steam output. It is very probable that the enlightened attitude in Italy towards the development of geothermal energy is the primary reason why the country is so far ahead. The overall policy is to maintain primary energy sources in the new operating steam fields (i.e. incidentally to compensate for wells long in use and showing decrease in flow rate by introducing new wells) and to exploit adjacent regions near known steam fields in order to increase overall steam production. Of course, there is also the continuing research and development programme. Additionally, there are the engineering factors, for instance developing technology to obtain the most suitable operating pressure and also the ever-widening employment of small turboalternator exhausting to atmosphere-type generating units for immediate utilisation of steam from new peripherally located wells. Finally, there are improvements in steam pipeline networks (including insulation) and the decentralisation of power stations (this latter reducing plant operating costs through the installation of remote control systems and also reducing pipeline costs and hence thermodynamic losses).

REFERENCES

1. Special Report (1973). *Ground Water and the Geothermal Resource.* Geraghty and Miller Inc., Water Res. Bdg, Manhasset Isle, Port Washington, New York, 14 pp.
2. Panichi, C., Ferrara, G. C. and Gonfiantini, R. (1977). Isotope geothermometry in the Larderello geothermal field. *Geothermics*, **5**, 81–8.
3. Panichi, C. and Tongiorgi, E. (1975). Carbon isotopic composition of CO_2 from springs, fumaroles, mofettes and travertines of Central and South Italy. *Proc. 2nd UN Symp. Development and Utilization of Geothermal Resources, San Francisco, 1975.*
4. Hulston, J. R. (1977). Isotope work applied to geothermal systems at the Institute of Nuclear Sciences, New Zealand. *Geothermics*, **5**, 89–96.
5. Bottinga, Y. (1969). Calculated fractionation factors for carbon and hydrogen isotope exchange in the system calcite–carbon dioxide–graphite–methane–hydrogen–water vapour. *Geochim. Cosmochim. Acta*, **33**, 49–64.
6. Gunter, B. D. and Musgrave, B. C. (1971). New evidence on the origin of methane in hydrothermal gases. *Geochim. Cosmochim. Acta*, **35**, 113–18.
7. Ferrara, G. C., Panichi, C. and Stefani, G. (1970). Remarks on the geothermal phenomenon in an intensively exploited field. Results of an experimental well. *Proc. UN Symp. Development and Utilization of Geothermal Resources. Geothermics, Sp. Issue*, **2**(2), 578–86.

166   GEOTHERMAL RESOURCES

8. Bottinga, Y. (1968). Calculation of fractionation factors for carbon and oxygen isotope exchange in the system calcite–carbon dioxide–water. *J. Phys. Chem.*, **72**, 800–8.

9. Del Pezzo, E., Guerra, I., Luongo, G. and Scarpa, R. (1975). Seismic noise measurements in the Monte Amiata geothermal area, Italy. *Geothermics*, **4**, 40–3.

10. Luongo, G. and Rapolla, A. (1973). Seismic noise in Lipari and Vulcano Islands, Southern Tyrrhenian Sea, Italy. *Geothermics*, **2**, 29–31.

11. Cappello, P., Lo Bascio, A. and Luongo, G. (1974). Seismic noise survey at Solfatara Crater, Phlegraean Fields, Italy. *Geothermics*, **3**, 76–80.

12. Palumbo, A. (1966). Osservazioni geotermiche alla Solfatara di Pozzuoli. *Boll. Soc. Nat. Napoli*, 75.

13. Rapolla, A. (1974). Natural electric field survey in three southern Italy geothermal areas. *Geothermics*, **3**, 118–21.

CHAPTER 12

Geothermal Energy and the Environment

Although geothermal energy has advantages over other industrial operations in that it is, so to speak, a cleaner method of producing energy, nevertheless there is bound to be some impact upon the environment and it is with this that the present discussion is concerned. Clearly, effects will be produced by such diverse activities as laying roads, drilling wells, installing pipelines and erecting power plants—all factors in the commercial exploitation of a geothermal resource. Such effects are going to have a great influence upon land utilisation patterns in any particular region. The exact extent of this influence will depend on the type of fluid and utilisation involved. When all this is taken into account, however, there is still no doubt that the production of electrical power from a geothermal plant is much less deleterious to the environment than its production from other types of thermal power plants or, in many cases, from hydroelectric plants as well. The latter may well have a dislocatory effect due to the enormous construction works involved in building them. In geothermal power production, every phase of the fuel cycle is localised at the site. Other kinds of thermal power plants need very considerable industrial back-up such as mines, transportation facilities and processing plants. All this enlarges the area of environmental impact of the fuel cycle for these operations a long way from the boundary of the actual power generating plant itself. In sum, therefore, it is possible categorically to state that geothermal energy is a relatively clean energy source because it does not pollute on generation like conventional fossil-fuel or nuclear plants.

Some environmental effects which are highly undesirable and produced by other operations, but *not* by those involved in producing geothermal energy, are those caused by refining, strip mining, industrial wastage, off-shore drilling, radioactive hazards of one kind or another and the necessity of transporting other kinds of energy. It may be useful now to look at the experience gained in connection with the vapour-intensive type of geothermal power plant which, of course, has a long history of production at Larderello, The Geysers and Matsukawa. In all three regions, the release of H_2S gas has been the major degradatory factor in the environment. Until now, this has not been considered particularly serious because all three are rather remote and not especially large as well as being in areas where natural H_2S is being released into the atmosphere anyway. As both the size and number of plants increase, however, the problem is likely to become acuter. Hydrogen sulphide is a peculiarly offensive pollutant which is not only obnoxious to everyone, but also all too easily detected. Here, the SO_2 emitted by coal-fired plants is by comparison not so bad. From geothermal plants, very little sulphur is actually emitted. The Environmental Protection Agency of the US Government stipulates that release of sulphur from fossil-fuel plants may not exceed 1·2 lb per million British thermal units and The Geysers emit less than a quarter of this. At The Geysers, the Pacific Gas and Electric Company is carrying out an emission abatement programme which is intended to remove 90 % of the H_2S from the non-condensable gases and the necessary equipment is to be included with new installations and added to old ones. Of course, the main gaseous release from geothermal plants is CO_2 and, as with sulphur, the amount released per unit of power is considerably below that emitted by fossil-fuel plants. It may be noted also that the geothermal plant will not liberate any smoke or oxides of nitrogen.

On the other hand, steam, contaminated water, noise and even seismic disturbances can spring from geothermal exploitation and constitute environmental pollutants. Visual pollution may result from the laying of unaesthetic pipelines which can offend the eye, but every attempt is made to minimise their effect; for instance, they do not obtrude at Larderello and actually exist in an agricultural region characterised by the growing of fruit and vegetables. Naturally, wells and power plants are not so easy to camouflage, but their presence need not interfere with other activities. An example is The Geysers where unused land is employed both as a deer reserve and for cattle grazing. Here, the wilderness aspect which formerly occurred is preserved. Of course, the large number of wells drilled—150 or more over a 12-square-mile area—is quite characteristic for a geothermal field and indeed additional acreage must be reserved for the later introduction of new wells aimed at maintaining the necessary steam supply. At Larderello also, the geothermal field has been made compatible with many other land uses during the 65 years and more of its development and production. The wells, gathering lines and power plant utilise only a set of small patches and strips of land, hence most of the land is being employed for a varied agricultural industry including a number of farms, vineyards and orchards actually interspersed among the pipelines and wells.[1] Noise has been referred to already and it does constitute an irritant. It is primarily associated with drilling operations and the escape of steam during testing. Once the geothermal field comes into production, the level of noise will abate and probably not exceed or even equal that emitted from other types of power plant. In fact, mufflers are in use at The Geysers during such drilling and testing operations and they have been found to be singularly effective. As regards seismic disturbance, the potential effects of this and also those which may arise from land subsidence resulting from geothermal development were discussed by the US Department of the Interior in an Environmental Impact Statement for the Geothermal Leasing Program (1972).[2] Fortunately, to date, neither seismic disturbance nor subsidence has been observed at The Geysers or for that matter at Larderello either. This may well be due to the fact that the geological framework of origination of dry-steam fields such as those mentioned is not conducive to subsidence. The production reservoir of a dry-steam geothermal field is fracturated, often very heavily, but there is a nearly constant vapour pressure observed vertically through the reservoir. For instance, at The Geysers, steam temperature and pressure are of the order of 240 °C and 34 kg/cm^2. The near constancy of pressure is manifest at great depth (over 8000 ft) where hydrostatic pressures would normally be approximately 280 kg/cm^2, and indicate that the rocks must be competent in order that the dry-steam field can exist at all. Hence, the removal of vapour will not trigger subsidence. It is very important to point out immediately that this is not true where hot-water fields are involved because these could act in a manner resembling an unconsolidated petroleum reservoir and subsidence might well be a possibility if the pressures are not adequately maintained by fluid return. That this is not merely a theoretical hazard is demonstrated by the fact that subsidence occurred at Wairakei in New Zealand where water was not returned to the reservoir, cf. J. W. Hatton.[3] Turning now to the seismic question, it must be remembered that where geothermal resources are located, unstable planetary crustal conditions operate and it is just these which also produce extensive faulting as well as earthquakes. In other words, seismic and geothermal activity are very likely to take place in the same areas and it is precisely this fact which renders seismic noise incidence a potentially useful tool in geothermal exploration (v., for instance, the preceding chapter and also G. R. T. Clacy[4]). Of course, shocks associated with volcanicity (one source of geothermal heat) are going to be far smaller in magnitude than those related to fault-governed major crustal movements and also there is absolutely no evidence whatsoever that geothermal production in any way *increases* the seismicity of a region. In fact, the main reason why such a hazard might arise is considered to be, at least partially, because sometimes waste geothermal fluids are reinjected and, where waste disposal operations involving fluid injection have been effected, for instance in the Rocky Mountain Arsenal near Denver, Colorado, related seismic activity has been observed. This may have resulted from the widening and lubricating of previously existing fractures or from the extension of a pre-existing fracture pattern. In geothermal reservoirs, however, low pressures exist and reinjection of fluids simply maintains pre-existing pressures and thus is highly unlikely to cause any increase in seismicity. In other words, since reservoir pressure is below hydrostatic, no pumping is required for reinjection of fluids (the actual weight of the water involved creating enough head to enable re-entry to occur without it) and, additionally, the reinjected fluid migrates to the region of minimal pressure and hence practically obviates migration into other aquifers.

From the above discussion, it may be inferred that, although a geothermal power plant may have a greater effect on the land initially than other thermal plants do, this is offset by the fact that all its constituent parts are on one site only. Contrast this with, for instance, a nuclear power plant in which the thermal reactor and the facilities for the generation of power comprise merely a small part of the power cycle. The actual transporting of nuclear fuels is a potentially devastating environmental hazard and this applies particularly to the disposal of nuclear waste. High-energy radioactive wastes may require adequate and permanent insulation from accidental entry into the environment and this may well entail not only reserving a large amount of surface land, but also allocation of a vast subterranean storage facility. Fossil-fuel generating plants, of course, need a huge land area, taking into account mines, railway yards and handling facilities and, for instance, with coal it is necessary to accommodate its washing and transportation as well as dispose of fly ash and clinkers. Combustion problems also arise and these are present too with oil-fired and gas-fired thermal plants, though to a lesser degree. Geothermal plants for producing power appear more and more attractive as the others are considered, but can only be erected where the geothermal resource occurs, naturally. One of the possible hazards arising from geothermal power development which must be mentioned, however, is the potential for the contamination of surficial and ground water.

12.1. THE POTENTIAL RISKS TO GROUND-WATER AQUIFERS

A classic instance of the potential risks to ground-water aquifers is afforded by the wells drilled into the Salton Sea in California, where the highly saline brines could be a bad contaminant if permitted to mix with irrigation waters in the area. Although geothermal waters are usually much less saline, nevertheless they are more saline than non-thermal waters and the possible contaminational effect remains to a lesser extent. However, thermal waters are occasionally pure enough to enable their employment in agriculture and industry. This means that their higher temperatures which promote the rate of solution of more volatile chemicals in the host rocks have not caused sufficient dissolution to add enough salts to them to constitute a hazard. N. V. Peterson and E. A. Groh have indicated that geothermal waters are utilised directly for stock watering at Klamath Falls, Oregon.[5] At this location, hot-water wells are drilled in a city and this has also been effected in Boise, Idaho, Budapest, Hungary and Rotorua, New Zealand. This would not be possible if a dry-steam well were to be involved. Dry-steam fields, for instance The Geysers and Larderello, do not present any problems in regard to saline-water disposal as a result of the fact that salts are not transported in the steam phase, most of the extraneous material being non-condensable gases. However, there is a very small amount of ammonia and boron present (a few parts per million) which can form salts which can stay in the condensate; they are later reinjected into the reservoir together with surplus condensed cooling water. The risks arising from dry-steam geothermal operations relate to development, a period when drilling muds are used and the usual water pattern of the region is dislocated by construction. These are not considered to be very important because they are transient and quite negligible by comparison with the hazards occasioned by competing power sources, for instance mining.

12.2. SURFICIAL WATERS

Surficial waters may be adversely affected by geothermal development activities. Over half of the total energy produced in a steam cycle is rejected as a consequence of thermodynamic considerations and this is usually effected by circulating cooling waters through a condenser in order to acquire reject heat and dissipate it in a suitable larger water body, often a

river or a lake. Clearly, the sudden arrival of this heat can trigger biotal
changes through thermal pollution. Consequently, it has become the
custom to dissipate such heat into the atmosphere rather than into
environmental waters and thus obviate the danger. Actually, large amounts
of water are necessary and its inadequacy may be a limiting factor in siting
thermal-generating plants. The significance of the problem is revealed by a
study which estimated that by 1980 a sixth of the fresh-water runoff in the
USA will be employed solely to cool power plants.[6] Of course, geothermal
plants evaporate more water than nuclear or fossil-fuel plants, being
thermally less efficient (about 15% for geothermal plants, cf. 33% for
nuclear plants and 38% for fossil-fuel plants). But this does not comprise
the entire picture since, for instance, the thermal efficiency of a nuclear-fuel
cycle is not just conversion of fission to steam energy—waste disposal alone
will use a lot of energy! Also, geothermal plants do not need any
supplementary sources of cooling water when using natural steam. This,
after passage through the turbine, is condensed and piped to the cooling
towers, thereafter being recirculated back to cool the condenser. In fact,
this method at The Geysers enables production of about one-fifth *more*
condensate than is evaporated and the return of the surplus to the reservoir
prolongs the working life of the field.

Harmful effects of contaminated waters on surface waters into which
they have been discharged have been noted at Wairakei where fish
downstream of the plant have been damaged.[7] This emphasises the
importance of heat dissipation into the atmosphere and removal of harmful
constituents.

12.3.   THE IMPACT ON THE ATMOSPHERE

Geothermal plants operate without combustion and it is hardly surprising,
therefore, that effects on the atmosphere are ridiculously small in
comparison with the impact produced by conventional fossil-fuel and
nuclear plants. Anyhow, most of them are quite remote and thus, for
instance, emission of H_2S at The Geysers has not been a cause of concern. It
constitutes no more than 225 ppm of the steam. As regards CO_2, it has been
shown that a fossil-fuel plant produces orders of magnitude more than a
geothermal plant for the same electrical capacity. The question of the
radioactivity of the gases and steam may be raised and this has been found
to be very near, if not actually at, natural background levels. For example,
steam from The Geysers has an alpha radiation level of $0.015 \times$

10^{-7} mCi/ml, i.e. well below the US Public Health Service permissible concentration for drinking water.[8]

Curiously, although a nuclear plant releases much less into the air, the end-impact from the *total* nuclear-fuel cycle is much greater than that of a geothermal plant. Initial mining operations release radionuclides as well as smoke and every step in the cycle is responsible for new releases. The sum is important, especially that part representing release from the fuel reprocessing plants.

Perhaps the best illustration of the relative cleanliness of the geothermal resource exploitation is given by Iceland where, in the 1940s, the town of Reyhaukur suffered such heavy smoke pollution that it was hardly visible to the inhabitants. Subsequent conversion to geothermal energy has resulted in clear skies and smokeless air.

12.4. SOME INFERENCES

It appears that the effect on the environment of any system of producing power is related to the complexity of the fuel and production cycle and here, geothermal power plants have a great advantage because they use naturally occurring steam and do not require complicated equipment to generate this. Nor do they need mining, processing or transportation facilities as do other thermal power plants. It appears also that the *main* potentially deleterious effects arising from the utilisation of geothermal power occur at the inception of the development of a field, i.e. the period of construction of lines for gathering steam and also power plants. However, these effects are of very limited scope and do not involve the terrible scarring of the landscape concomitant with strip mining, for instance. The anticipated life span of a geothermal field is certainly several decades, if not several generations, and during this period it is feasible to carry on other activities in the same region as at Larderello. The tiny percentage of non-condensable gases found in natural steam constitutes an environmental hazard of negligible proportions when compared with the vastly greater gas release and toxicity produced from the nuclear-fuel cycle, for instance. There is no hazard to water supply from dry-steam sources and, while hot-water geothermal systems do exert an effect on waters, this is usually to employ waters derived from depths below those normally economically drillable. Sometimes, too, they can improve the quality of waters considered unusable before the geothermal development took place. Finally, the fact that the geothermal steam cycle is independent of external sources of

174   GEOTHERMAL RESOURCES

electricity, needs no mining, railways or processing facilities and can be
operated by relatively few personnel militates strongly in its favour (where it
can be employed) as an extremely reliable source of energy.

REFERENCES

1. Bowen, Richard G. (1973). Environmental impact of geothermal development.
   In: *Geothermal Energy*, ed. Paul Kruger and Carel Otte. Stanford University
   Press, Stanford, Ca., pp. 197–215.
2. US Department of the Interior (1972). Supplement to draft: *Environmental
   Impact Statement for the Geothermal Leasing Program*. Revised Chapter IV,
   Sec. C—alternatives to proposed action; Appendix G—energy alternatives;
   Appendix H—proposed unit plant regulations, 175 pp.
3. Hatton, J. W. (1970). Ground subsidence of a geothermal field during
   exploitation. In: *UN Symp. Util. of Geothermal Resources, Pisa, Italy*. Maxwell
   Scientific International, New York.
4. Clacy, G. R. T. (1968). Geothermal ground noise amplitude and frequency
   spectra in the New Zealand volcanic region. *J. Geophys. Res.*, **73**, 5377.
5. Peterson, N. V. and Groh, E. A. (1967). Geothermal potential of the Klamath
   Falls area, Oregon; a preliminary study. *Ore Bin*, **29**(11), 209–31.
6. Holcomb, R. W. (1970). Power generation: the next 30 years. *Science*, **167**,
   159–60.
7. Sullivan, W. (1975). 'Analyst warns of pollution from geothermal projects.' In:
   New York Times, March 1 issue.
8. Bruce, A. W. and Albritton, B. C. (1959). Power from geothermal steam at The
   Geysers power plant. In: *Proc. Amer. Soc. Civ. Eng., J. Power Div.*, **85**, PO6,
   Part 1, 34.

CHAPTER 13

Research into Geothermal Resources: Some Interesting Recent Developments

The most cursory examination of Man's history will demonstrate that awareness of the geothermal resource has existed for millennia, ever since, in fact, our ancestors noted the occurrence of and products from volcanoes, geysers, fumaroles, mofettes, warm and hot springs and mud pools containing boiling waters. In addition, drilling into the planetary crust has been going on for over a hundred years, initially for petroleum, and, for almost half a century, wells have been made in deep water conditions. It is only in the past quarter century or so, however, that these activities have been directed specifically at exploiting the geothermal resource for the production of power and this is because knowledge of this has been rather sparse. It is in order to provide more essential data that the US Government and other authorities have encouraged research into all aspects of the geothermal question, but the sums of money which have been allocated for this purpose have not been anywhere near adequate. For instance, Jesse C. Denton and Donald D. Dunlop indicated that in the fiscal year 1973, a mere $US6·5 million was earmarked for identified geothermal programmes of various Federal US agencies.[1] Obviously, a lot more will have to be forthcoming if development is satisfactorily to proceed. It will be advisable here to examine first some aspects of the problem and later some fields in which notable progress may be expected, not only in the USA but also elsewhere. Of course, the two other countries with optimal possibilities are New Zealand and Italy, but some others, for instance Iran, offer an interesting future potential.

13.1.   THE EXPLORATION PROGRAMME

Clearly, the major aim of exploration will be to identify regions containing heat, that basic commodity with which society is concerned. Hot rock with a suitable cap rock is what is needed. It is also necessary to determine the total volume involved as well as the temperature situation and also the permeability. The existence of fluids and their nature are other very important factors about which information is required. As was seen earlier in this book, there are a number of different types of geothermal resource which may be divided conveniently into four categories, namely convective hydrothermal, geopressured, hot impermeable rock and magma systems. Nobody has been able yet to exploit molten igneous materials, although, as indicated in Chapter 2, H. C. Hardee and Sandia Laboratories are investigating possibilities of extracting thermal energy from them.[2] Any investigation of the types mentioned must take account of the fact that there is a considerable difference between them and indeed each one has its own specific physico-chemical properties. With convective hydrothermal systems, it is important to determine the relationship with the responsible magma body and also the size, age and kind of this. Naturally, the structural controls on the location of such systems as well as hot and impermeable rock systems must also be examined. As noted earlier in Chapter 3, the accumulation of geochemical and geophysical data is all-important in every stage of a feasibility study as well as in later activities, where these take place, devoted to development and utilisation. Intensive study of the physico-chemical and thermodynamic properties of aqueous solutions at 'geothermal' temperatures and particularly also of the isotopic variations of 'geothermal' waters is a must.

There is great variation shown by geothermal systems in electrical resistivity and efforts must be made to assess better the effects of such parameters as porosity and salinity and temperature upon this in geothermal reservoirs. All possible forms of electric exploration really require more research, for instance de-resistivity, telluric and magnetotelluric. They have an advantage in that they can comprise a part of an airborne survey. Some aspects of electrical methodology have been already discussed in Chapter 11 (section 11.3).

Seismic technology is also useful and it may be useful to determine the patterns of earthquakes more precisely in appropriate regions because this can assist in the finding and mapping of faults which may conduct hot fluids to depths to which Man can drill using existing technology. The importance of seismic noise studies has also been indicated earlier in Chapter 11 (v. section 11.1.3 and section 11.2). They can evaluate temporal-spatial noise variations and their causes, the direction of noise propagation and apparent noise velocity as well as contributing to an understanding of the occurrence of geothermal areas.

Gravimetric and magnetic work is also invaluable because it shows that these properties vary greatly from one geological province to another. However, more research is necessary in order to comprehend better the *sources* of anomalies associated with geothermal areas and also to show how far such anomalies can be of use in the determination of subterranean temperatures, if indeed they can be of any use at all in this respect.

There is also a strong possibility that the construction of suitable laboratory models of geothermal systems might prove to be valuable in the study of temperatures beneath the crustal surface and, of course, hydrologic modelling of the conventional kind may well aid greatly in understanding better the effects of ground-water movement on the local geothermal gradient.

Clearly, more advances in drilling technology would be tremendously beneficial. This especially relates to drilling to depths greater than those hitherto attained and it would be excellent if depths in excess of 8–10 km could be reached. Problems arise, of course, from high temperatures which are deleterious to seals, valves, cements, drilling muds, mufflers, etc. Logging devices capable of employment at temperatures above 200 °C are also a basic necessity. Rotary drills need improvement and among new approaches suggested are erosion drills, electric-melting drills and turbine drills.[1]

Once these advances in basic methodology and computer modelling have been achieved, the subsequent evaluation of the geothermal resource may be carried out more efficiently.

13.2. ASSESSMENT OF THE GEOTHERMAL RESOURCE

The geothermal resource normally comprises not only heat, but also water and minerals. Obviously, major emphasis is placed upon heat, i.e. energy, but the other components are also important and useful.

Great variation occurs between estimates of resources by different authorities and this indicates the urgent need to improve evaluation techniques. This is quite startlingly indicated by comparing Donald E. White's 1965 statement that between 5000 and 10 000 MW of power could be generated for a minimum of half a century under existing economic conditions and technology with that of the National Petroleum Council (USA) in 1972 that by 1985, the United States could be generating 7000–19 000 MW from the geothermal resource.[3,4] Also, R. W. Rex believed (in 1972) that the USA could develop an amazing 400 000-MW capacity within two decades and that this capacity could be kept up for a century.[5]

Of the types of geothermal resources mentioned earlier, the magmatic one may be left out of account momentarily. As will be recalled, the hydrothermal convective system comprises vapour-dominated systems and liquid-dominated systems and the former are rare, although they account for most *existing* geothermal power production on Earth. With both, it is necessary to find out the size and life history of the resource, and also all other relevant parameters in order to make a suitable engineering development of future fields (which will be probably liquid-dominated) feasible. Geopressurised aquifers are to be found in fault-bounded zones which in the USA run parallel to the Gulf coastline. They yield heat and water and also mechanical energy from high pressure and contained methane. Obviously, they should be researched, in this case into the continental shelf.

13.2.1. Hot-water Heat-exchanging Systems

Flash production of steam is not practical at hot-water temperatures, i.e. between 100 °C and 200 °C, because steam flashed from geothermal hot water is corrosive and also because of high water consumption and low cycle efficiencies. Additionally, the steam turbines become more expensive handling low-temperature steam.

The answer is to utilise some appropriate secondary fluid by passing the hot water directly through a heat exchanger and imparting its thermal energy to freon, isobutane or some fluorinated hydrocarbon. There are many advantages to such a system and these include the following:

1. smaller turbine sites;
2. reduced corrosion potential—isobutane, for instance, is almost non-corrosive;
3. an isobutane turbine operates above atmospheric pressure and this positive pressure militates against the intrusion of air and thus again reduces the corrosion potential;
4. there is less stress and vibration in the turbine;
5. turbine efficiency is improved.

The major disadvantages are, of course, that

1. a heat exchanger costs money;

2.   secondary fluids of the type indicated above are more expensive
      than water and therefore leakage from the closed system has to be
      entirely eliminated.

13.2.2. Dry Geothermal Source Systems

Practical development of dry geothermal systems is very important
because, in point of fact, nearly all the unused geothermal heat resources of
the planet are situated in dry, hot rock sources. A project aimed at studying
the feasibility of developing means of extracting thermal energy from these
was commenced at the famous Los Alamos Scientific Laboratory in 1975
(v. Fig. 13.1).[6] This means, of course, heat-energy extraction on a
commercial scale.

The area selected is Valles Caldera located in northcentral New Mexico,

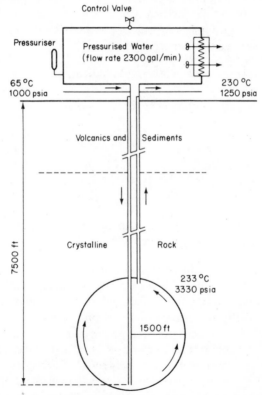

FIG. 13.1.   Experiment at Los Alamos involving a dry geothermal source (hot rock
system).

USA. Here, the most recent volcanic activity is thought to have occurred about 50 000 years ago and it was concluded that a great deal of heat remains in the rocks lying not far below the surface of the crust. In fact, this is demonstrated by the presence of hot springs near the western periphery of the caldera. In 1970, an informal group of Los Alamos Scientific Laboratory staff began to investigate possibilities of developing a rock-melting drill (Subterrene) which had been invented some 5 years earlier. In 1971, the Los Alamos Scientific Laboratory Geothermal Energy Group was formed and made heat-flow measurements around the caldera. Maximal values were found to the west of this in a region where the geology turned out to be least disturbed and there is only a moderate depth to the Precambrian basement. Thereafter, in 1972, a 2576-ft-deep hole was drilled in Barley Canyon, 5 km west of the caldera and this was called GT-1. In 1973, the Geothermal Energy Group became re-designated Group Q-22 and obtained USAEC funding which facilitated the drilling of a second hole, GT-2. A third hole was projected to attain 6701 ft in 1974. As regards GT-2, the site selection criteria were

1. an over-average geothermal gradient,
2. simple geology and structure,
3. near-surficial competent rocks for a heat reservoir.

During the actual drilling operation, at 1961 ft the drilling fluid was lost and thereafter air was used. The third hole actually attained 6701 ft and a temperature of 146 °C was recorded, the temperature gradient being calculated as about 50 °C per 1000 m. A dual project plan was set out for 1975 and this involved either

1. drilling of a first energy experimental hole to be called EE-1, or
2. effecting more experiments in GT-2.

Figure 13.1 gives details of the GT-2 experimental programme.

A hydraulic fracture network is to be made with a diameter of approximately 3000 m. Subsequent to the initial pumping of water into this (which is actually an underground heat sink), the system is to operate as a closed loop with compensatory water to be added to replace losses. Water at 65 °C and 1000 psia is to be pumped into the earth and circulated at 7500 ft where the temperature range is between 260 and 320 °C and pressures around 4100 psia are anticipated. Water will then be pumped out to reach the surface with a temperature of 230 °C at 1250 psia. For the experiment, hot water is circulated through a 100-MW air-cooled heat exchanger with extracted heat dissipated to the atmosphere. Of course, a lot of testing is necessary before *commercial* feasibility can be established. Probably, new thermal fracturation must be introduced from time to time in order to enlarge the heat-transfer surface area available. However, it does not matter whether the system shown in the diagram is eventually used in industry or a simpler one adopted because either will yield hot water, the object of the exercise. If this is sufficiently hot, steam might be flashed from it in order to drive a steam turbine. With cooler water, this could be passed through a heat exchanger in order to operate a turbine through a secondary drive fluid. One of the great advantages would be the totally non-polluting aspect of such a practically totally enclosed geothermal system.

13.3. THE COMBINATION OF RESOURCES

Obviously, the provision of a hot fluid at the surface of the Earth is a prerequisite of power generation and, as noted, this can derive either from that of the reservoir involved or it may be the result of injection from the crustal surface. Either way, minerals may also be produced as a result of the fact that hot water, for instance, is a first-class solvent for many of them.

The actual engineering involved depends upon the assessments made of size and also *deliverability* of the geothermal resource as well as adequate planning. This is admittedly a pejorative word, but, if properly carried out, can be an absolutely key factor! Well-stimulation techniques, cf. Chapter 5, can improve the permeability of a formation and thus both production and reinjection. The economics of the production of electricity geothermally justifies the multiple-use approach in order simultaneously to reduce expenses and increase efficiency. There are two processes which are applicable, namely an integrated one and a so-called divorced one (Fig. 13.2).[7]

13.3.1. Integrated System

The integrated system involves wet-steam brine being transmitted through a wellhead separator, the steam and brine then passing through a desalination plant where

1.   water is separated from brine, and
2.   steam is purified, i.e. H_2S, CO_2, etc., are removed.

The cleansed steam is utilised to drive the turbine for the production of power and the condensing steam from this is combined with the desalinated

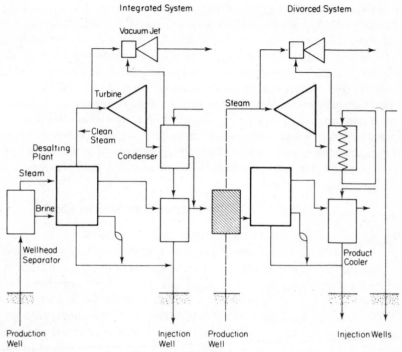

FIG. 13.2.   Production of water and electricity from one power plant: alternative
processes.

water in the cooler in order to produce pure water which can, of course, find
application in agriculture and industry as well as, for that matter, in
domestic outlets. Residual brine is mixed with compensatory water and
subsequently reinjected into the ground.

13.3.2. The Divorced System

The sole distinction of the divorced system is that the individual electrical
and water production facilities are separated.[8]

13.4.   RESEARCH AND DEVELOPMENT IN NEW ZEALAND

So far, research and development in the USA alone have been discussed,
but it may be useful to consider recent developments in New Zealand,
another, if smaller, world 'geothermal power'.

J. R. Hulston has given an excellent account of isotopic applications.[9] Actually, isotopic investigations on geothermal areas began as early as the middle 1950s, almost immediately after work on the geothermal areas. Initially, it was considered appropriate to concentrate on sulphur and carbon and on dating using radiocarbon (cf. Chapter 3). Later, hydrogen and oxygen isotope measurements on water, and oxygen isotope measurements on rocks and sulphates, as well as hydrogen isotope measurements of methane and hydrogen gases, were actually carried out. Isotopic geothermometry was a primary aim of some of these researches. One of the first pieces of work was effected by the group on the sulphur isotopic composition of H_2S from the Wairakei region and calculated temperatures of around 370 °C agreed rather badly with measured temperatures in the wells of a maximal 270 °C. An uncertainty existed as to whether isotopic equilibrium obtained or not. Later work on carbon suggests that it does. Hulston certainly believed this, but the matter is still, to a degree, *sub judice*. However, bisulphate-water, oxygen-18 and H_2-water temperatures have been found to be in fairly good concordance with those obtained from the Na–K and silica geothermometers which are now frequently employed elsewhere and really comprise routine tools. Studies have been made upon the distribution of sulphur isotopes between mineral phases in the famous Broadlands geothermal field (cf. Chapter 7, section 7.2) and it has been found that, under the chemical conditions of this system, equilibrium between sphalerite and galena is achieved above 250 °C. In this same field, oxygen isotopic equilibrium was found to exist between fissure-grown quartz, adularia (a low-temperature potassium feldspar) and calcite at temperatures of 250–290 °C. Also, there appears to be oxygen isotope equilibrium between secondary quartz and illite, a clay mineral, at temperatures of 260–270 °C. Interestingly, these data are in good agreement with those derived from the Salton Sea area in California where oxygen isotope equilibrium between fine-grained silicates and water is achieved above 150 °C. At Wairakei, at temperatures of 250 °C, oxygen isotope equilibrium appears to occur in whole rock samples in zones of geothermal alteration.

As regards dating, early data showed ^{14}C activities in the range 0·4–1·8 % of normal ground water and no doubt these low levels reflect addition of old CO_2 to the system either from magmatic gases or from the in-depth decomposition of carbonates. The extent of this dilution constitutes a problem and investigations have been made into the feasibility of using the atmospheric argon/carbon-14 ratio as a method of estimation of this.

Tritium has also been employed at Wairakei, actually on well discharges over a long period, more than 15 years. In the beginning of the development of the field, the values obtained were rather on the low side. It is thought that the presence of this radioisotope at all showed shallow recirculation in the system. Measurements made in 1960 gave results from 0·04 TU to 0·44 TU and analyses completed in 1972 yielded results from 0·14 TU to 0·5 TU, i.e. negligible changes are displayed and in any case, the writer would consider the amounts to be too small to be significant. This would suggest that the deeper ascending waters are most likely tritium-dead.

Some work has also been done on the question of isotopic compositions and element origins.[9] The carbon-13/carbon-12 ratios of carbon gases from the New Zealand geothermal areas gave values usually within the range from $\delta^{13}C_{PDB} = -3\%_0$ to $-7\%_0$, quite similar to the Yellowstone Park data of H. Craig.[10] Although juvenile carbon cannot be altogether excluded, nevertheless it does appear possible to explain this composition from mixed carbonate–organic carbon within the rocks themselves. It has been suggested that sulphur isotopes from sulphate in the well discharges at Wairakei may be of sea-water origin, sulphide having isotopic ratios similar to that for meteoric sulphur deriving from a magmatic source. However, recent work shows that both sulphide *and* sulphate of practically zero $\delta^{34}S$ values could be derived from basement sedimentary rock. Probably, the sulphur in New Zealand thermal waters has a mixed sedimentary and magmatic origin.

13.4.1. Ore Bodies: Their Influence as Indicators of Past Geothermal Conditions

In the Coromandel Tertiary volcanics area, a stable isotope study is in progress in regard to base-metal, gold and silver-bearing lodes of quartz. This lies northwest of the principal geothermal region of New Zealand. The $\delta^{13}C$ value of total carbon and the $\delta^{34}S$ value of total sulphur show a crustal origin just as in the current geothermal region. The isotopic composition of the carbonates becomes heavier with time (demonstrating decreasing temperatures). Isotopic analyses of fluid inclusions showed that the ore fluid was initially magmatic in derivation, but, after faulting occurred and fluid circulation increased in this region, so a geothermal system began and *later* ore fluids comprised deeply circulating meteoric water.[9] The δD values of these later waters are heavier than the meteoric water of today and most likely indicate a warmer Tertiary climatic regimen. The Tertiary volcanic mineralisation event is dated at between 2·5 and 7 million years before the present by potassium–argon methods.[11]

13.4.2. The Origins and Circulation of Water from New Zealand and Other Countries

It is very interesting to note that, according to J. R. Hulston, hydrogen and oxygen isotopic compositional studies at the Institute of Nuclear Sciences, Lower Hutt, New Zealand, effected on geothermal waters not only from New Zealand but also from Indonesia, the Philippines, Japan, Kenya, Chile and Antarctica demonstrated that generally they have a local meteoric origin, but underwent varying oxygen shifts as a consequence of reaction with deep-seated rocks at the temperature of the relevant systems.[9] Clearly, it is important to sample, accurately and systematically, two-phase flows and the output characteristics of wells must be understood if the compositions of the subterranean fluid are to be derived. Harmon Craig in 1963 stated that there was practically no oxygen shift at Wairakei, but R. N. Clayton and A. Steiner later (1975) determined values of $\delta^{18}O$ of -5.5% to -6.3% for the deep fluid.[12,13] Also, these workers noted that altered rocks in the field had an average shift of -4% away from the unaltered ones of the same type. They used $+0.5\%$ as the *upper* limit of the oxygen shift of the water and calculated a lower limit for the water/rock mass ratio of 4·3, roughly greater by a factor of 10 than that calculated for the Salton Sea geothermal region.[14] By contrast, Hulston cited a Wairakei local meteoric water ^{18}O value as high as -6.7%, implying an oxygen shift there of more than 1% and a water/rock ratio of 2 at maximum.[9] There is believed to be a somewhat greater shift at Broadlands. However, steam loss from water at 250 °C can produce an apparently smaller shift for ^{18}O and a larger one for steam and there has been suggested a variable H_2S loss at Wairakei which could imply a loss of steam.[9] Obviously, it is very important to correlate isotope data with chemical data.

Turning to Chile, isotope study of waters from El Tatio showed that this is a field where water is derived from outside the catchment area in which the geothermal field is located and the effects of dilution by surface water and steam separation are clearly apparent.[15]

An exciting application of a radioisotope tracer in order to monitor the natural directions of flow in the Broadlands geothermal field after all the wells had been closed in for several years has been made using iodine-131. This has a half-life of 8·05 days and is usually used in the iodide form. It emits low-energy gamma radiation (principal energy 0·36 MeV), is easily detectable and has a maximum permissible concentration of 3×10^{-6} $\mu Ci/ml$. At Broadlands, it was injected into selected wells and thereafter monitored from them and others. The apparent horizontal velocities were calculated at about 0·1 m/h.

13.5. TUSCANY

Interesting geothermal regions have also been located at Tuscany by application of geochemical and isotopic methods and reference may be made to work done there by R. Fancelli and S. Nuti.[16]

In western and central Siena province, i.e. an area lying to the east of the famous Tuscan geothermal fields to which allusion has already been made (v., for instance, Chapter 11), a number of zones have been individualised as being of interest with respect to geothermal fluid research. Utilisation of the Na–K–Ca ratio geothermometer has facilitated the delineation of those possessing high subterranean temperatures. Oxygen isotope analyses have demonstrated that exchange between the oxygen of the water and that of the host rocks has occurred there.

The inference drawn was that the geothermal manifestation (associated with basement fracturation) results from an admixing of geothermal steam condensate with shallow circulation waters.

An examination of the relevant geology and structure is important and reveals that the area under discussion is a wide tectonic depression which was formed during the Pliocene tensional phase and thereafter infilled with argillaceous (muddy) marine deposits which were affected by regional uplifting during the subsequent Quaternary. The maximal elevation achieved, actually in the Monte Amiata area, was about 900 m above mean sea level.

Peripherally, there occur small and isolated outcrops of the so-called 'Tuscan series' terrains belonging to three complexes, namely

1. basal (Permo–Triassic phyllites, quartzites and conglomerates);
2. calcareous, structurally discordant with respect to (1) above, Mesozoic anhydrites, dolomites and limestones;
3. argillaceous and arenaceous (sandy) limestones (Cretaceous to Eocene in age) together with cataclastic formations belonging to the allochthonous (transported) cover rock of flysch in the Ligurian facies (Lower Cretaceous to Miocene in age).

Hydrogeologically, the Mesozoic carbonates and fractured permeable rocks of the basal complex represent, on outcropping, excellent absorption areas which, therefore, determine the hydrostatic pressure. Where covered by a suitable cap rock, they constitute the main reservoir to which the hydrothermal manifestations are connected.

The surface manifestations include a number of hot springs which occur on dislocation lines associated with overfolding, and chemically these are characterised by the presence of SO_4 and Ca on the west and by HCO_3 and Na on the east. One of the best-known hydrothermal sites is that of the Petriolo Spa where the temperature ranges from 30 °C to 40 °C. Of course, all these phenomena relate to an anomalous heat flow, but this is true of the entire Tuscan province, i.e. a region bordered to the north by the River Arno, to the west by the Tyrrhenian Sea, to the east by the Chiana Valley and to the south by the valley of the Tiber. In the peripheral areas, gradient values have been recorded as 0·5 °C per 10 m, while centrally, they are quoted as 2 °C or more per 10 m and at Larderello, they rise to 5 °C per 10 m. Some conclusions may be drawn and these are as follows:

1. the region is heavily tectonised;
2. faults and transverse fractures are responsible for displaced blocks;
3. the structure is important as a fluid store and also for circulation, defining a number of subterranean water basins each having its own characteristics;
4. subterranean values of temperature have been evaluated.

Obviously, further drilling is necessary further to enhance the picture and render it more complete.

13.6. IRAN

Some allusions have already been made to Iran and its geothermal potential, and the writer regards it as one of the most promising of the so far unexploited regions of the planet. Four areas are involved: Maku–Khoy in the extreme northwest of the country, a zone bounded on the west by Turkey and on the north by the USSR; Sabalon and Sahand, two volcanoes to the southeast and not far from the famous carpet centre of Tabriz; and finally Mount Damavand northeast of the capital, Tehran (v. Fig. 13.3).

In 1978, the writer worked on the Mount Damavand area with Tehran Berkeley Civil and Environmental Engineers, actually in connection with the initiation of the second stage of a project to develop ultimately a 50-MW electrical power generating facility. The first stage of this work developed from an agreement between the Imperial Iranian Government's Ministry of Energy and ENEL, the Italian company which effected a reconnaissance survey starting in 1976. The precise details of the results are confidential, but it may be indicated that all four areas yielded very promising

FIG. 13.3.   Geothermal areas of Iran. Sites of major former earthquakes are shown and these lie within extensive areas of recent volcanicity extending into Turkey (cf. Mount Ararat) and demonstrate plate movements and the long-term effects of the Alpine orogeny. In the Maku–Khoy area, for instance, there is extensive fracturation and faulting while, to the southeast, the Alborz Mountains are of Middle Tertiary age and represent ancient Tethyan sediments. Mount Damavand is not extinct as are Sahand and Sabalon-e-Kuh, but fumarolically active. All the areas are characterised by the presence of many mineral and hot springs and thus evidently constitute potential geothermal resources. The occurrence of significant tritium in the ground-water system at Damavand also shows that a suitable aquifer and circulation system probably exist.

information. Damavand is the optimal one because of the following reasons:

1. it is the site of an active volcano with fumarolic activity (Sabalon and Sahand are extinct);
2. it is proximate to the capital;
3. quite a lot is known about the regional geology as a result of mapping work effected by a number of European geologists in fairly recent years and also by the United Nations;
4. it is the site, obviously, of a marked thermal anomaly;
5. for Iran, the access is not too difficult (compared, say with Maku–Khoy);

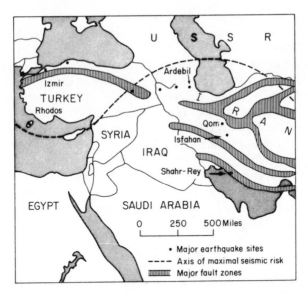

FIG. 13.4.   Maximal seismic risk axis for the Iranian Plateau and Middle East. Major fault zones are shown in this diagram together with major earthquake sites and these demonstrate the seismic instability of Iran which is related to the thrust northeastwards of the Arabian plate against the Eurasian one. The eastern Iranian area of the Kut Block, believed to be a stable region, appears to suffer as much seismic disturbance as anywhere else in the general region.

6.   many hot and mineral springs exist in the area which can be sampled and analysed;

7.   excellent cap rocks exist, for instance the Jurassic Shemshak (Lias) sandstones;

8.   tectonic activity has fractured many of the strata.

A number of observations may be made. The region is clearly one which was affected by the tremendous orogenic events of the Tertiary and forms part of the mighty Alborz range flanking northern Iran which, together with the Zagros range, constitutes still visible evidence of these historical geological occurrences. Site of the Tethys, Iran is still very unstable tectonically, for example one may cite the Tabhas earthquake of the fall of 1978 which killed thousands in the eastern part of the country and was felt by the writer in Tehran (v. Fig. 13.4). Every year, in fact, thousands of seismic shocks of varying degrees of intensity occur in Iran. As a consequence of all this, Mount Damavand formed and most probably very recently. If the volcano had existed at its current elevation in the Pleistocene, there would be evidence of glaciation on it. There is no such evidence and it may be inferred, therefore, that it reached this elevation *after* the glacial period, i.e. within a few thousand years of historic time. Observations show that there is an excellent circulation system for ground water underground and the waters have been shown to possess significant amounts of tritium, indicative of rapid transit time. The geothermal anomaly is believed to lie to the north of the mountain, i.e. on the Caspian side, and further studies, particularly by electrical methods, were planned. However, happenings in Iran in 1978 resulted not only in seismic, but also in political shocks, which, most unfortunately, aborted the intended programme—at least for a while. It is very much to be hoped that it can resume at a later and happier stage, for it would undoubtedly benefit the country.

REFERENCES

1. Denton, J. C. and Dunlop, D. D. (1973). Geothermal resources research. In: *Geothermal energy*, ed. Paul Kruger and Carel Otte. Stanford University Press, Stanford, Ca., pp. 335–46.
2. Hardee, H. C. (1974). *Natural Convection in a Spherical Cavity with Uniform Internal Heat Generation.* Sandia Laboratories, SLA-74-0089, 20 pp.
3. White, D. E. (1965). Geothermal energy. *US Geol. Surv. Circ.*, **519**.
4. National Petroleum Council (1972). *US Energy Outlook, An Interim Report.* Initial Appraisal, New Energy Forms Task Group, 1971–1985.
5. Rex, R. W. (1972). Testimony to Senate Interior Committee, June 15.
6. Smith, M. C. (1975). The Los Alamos Scientific Laboratory dry hot rock geothermal project (LASL Group Q-22). *Geothermics*, **4**, 27–39.
7. Cheremisinoff, Paul and Morresi, Angelo C. (1976). *Geothermal Energy Technology Assessment.* Technomic Publishing Co. Inc., Westport, Conn., 164 pp.
8. Porter, L. R. (1973). Geothermal source investigations. *Proc. Amer. Soc. Civil Engrs., J. Hydraulics Div.*, **99**, N.HY11, 2097–111.
9. Hulston, J. R. (1977). Isotope work applied to geothermal systems at the Institute of Nuclear Sciences, New Zealand. *Geothermics*, **5**, 89–96.
10. Craig, H. (1953). The geochemistry of stable carbon isotopes. *Geochim. Cosmochim. Acta*, **3**, 53–92.
11. Adams, C. J. D., Wodzicki, A. and Weissburg, B. G. (1974). K–Ar dating of hydrothermal alteration at Tui Mine, Te Aroha, New Zealand. *N.Z. J. Sci.*, **17**, 193–9.
12. Craig, H. (1963). The isotopic geochemistry of water and carbon in geothermal areas. In: *Nuclear Geology in Geothermal Areas*, ed. E. Tongiorgi, pp. 17–53. Consiglio Nazionale delle Ricerche, Laboratorio di Geologia Nucleare, Pisa.
13. Clayton, R. N. and Steiner, A. (1975). Oxygen isotope studies of the geothermal system at Wairakei, New Zealand. *Geochim. Cosmochim. Acta*, **39**, 1179–86.

14. Clayton, R. N., Muffler, L. J. P. and White, D. E. (1968). Oxygen isotope study of calcite and silicates of the River Ranch Number 1 well, Salton Sea geothermal field, California. *Amer. J. Sci.*, **266**, 968–79.

15. Cusicanqui, H., Mahon, W. A. J. and Ellis, A. J. (1975). The geochemistry of the El Tatio geothermal field, Northern Chile. *Proc. UN Geothermal Conf., San Francisco*, **1**, 703–12.

16. Fancelli, R. and Nuti, S. (1974). Locating interesting geothermal areas in the Tuscany region (Italy) by geochemical and isotopic methods. *Geothermics*, **3**, 146–52.

Note: There is a good geological survey of Iran by J. Stöcklin which he effected as a United Nations Expert and published through the Imperial Iranian Geological Survey in 1977, but it is practically impossible to obtain.

CHAPTER 14

The Future of Geothermal Energy

There still remain a few areas of doubt about the feasibility of geothermal power making a really sizeable contribution to the needs of those countries, notably the USA, Italy, New Zealand, Japan and Iceland, where its actuality and potential are significant. Obviously, new sources of energy are tremendously important because the classic fossil-fuel sources are being depleted at such an unfortunate rate. As has been noted, the natural heat of the Earth is geothermal energy, a vast resource. However, it must be added that another, even greater, natural resource exists and this is solar energy in all its aspects. In order to assess the future of geothermal energy, it is relevant to examine solar energy briefly and attempt a comparison.

14.1.   SOLAR ENERGY

The idea of exploiting solar energy is quite old, dating back at least 100 years and it is true to say that a number of recent proposals are mere re-presentations of the suggestions of earlier years with the addition of improved materials and engineering technology. The advent of the exploration of space has also provided new methodological approaches. In the early 1960s, a surge of interest arose in solar energy development exactly as with geothermal energy because of the concern then beginning to be felt regarding the pollution of the atmosphere and water as a result of the combustion of fossil fuels. In the early part of the current decade, anxiety over depletion of these emphasised the attractions of both solar and geothermal energy sources, attractions increasing because of the fantastic rise in charges for both crude and refined petroleum products consequent upon the Yom Kippur War between Israel and the Arab states in 1973.

192

A number of possible fields of solar energy technology exist and these may be summarised thus:

1. Solar climatic control: This is the term which may be utilised in order to describe solar radiation collection systems using appropriate energy-absorbent materials located in optimal positions such as the roofs or walls of buildings. To indicate the practicality of this, in the middle of 1975 there were no less than 120 *important* solar-heated structures world-wide. An extension is the heating of air rather than water.

2. Solar energy focusing systems: In these, the solar energy received over a sizeable area is focused on suitable energy-absorbing materials for later transport to a fluid medium, for instance water, to be used in a converting device ultimately to produce enough electrical power to conserve conventional fuels by electrical utilities. The methods under test include a solar tower energy collector and a solar absorption coating and heat pipe system.

3. Direct solar energy conversion systems: In these, thermionic and thermoelectric devices receive solar energy input and provide an output of electrical power.

4. Indirect solar energy conversion systems: Here, the solar energy is converted into a storable form of energy, for example hydrogen (through electrical or thermal dissociation of water into its components).

5. Ocean thermal energy conversion: The principle involved is the existence of a difference of temperature between the upper, sun-heated layer of the sea and the underlying, therefore deeper, cold-water layers. Clearly, the method would involve the use of off-shore platforms and specially designed turbines.

6. Wind power: Winds originate because of atmospheric thermal differences, hence their energy is attributable to the sun.

7. Photosynthesis: Sunlight could be used more efficiently in growing agricultural products. Progress in this and the other techniques has not been as rapid as could be wished, in part because of the competition from geothermal energy.

It may be of interest to discuss some of the more unusual techniques hereafter.

14.2.   OCEAN THERMAL ENERGY CONVERSION

The principle of the method for ocean thermal energy conversion was first indicated as long ago as 1881, and by 1930 a primitive ocean thermal difference power plant was in operation at the edge of Matanzas Bay in Cuba. This involved a 1·6-m diameter, 1·75-km-long, cold-water pipe from the on-shore plant into the bay where it reached a depth of 700 m. From a sea-water temperature difference of 14 °C, the turbine generated 22 kW of electrical power. As the designer, G. Claude pointed out, the plant operated as an open Rankine cycle system, i.e. steam generated from the sea water was utilised directly as the working fluid for the turbine.[1] Another such installation was made by the French off the Ivory Coast, Africa, in 1956.

14.3.   SATELLITE ENERGY SYSTEMS

Obviously, there is a great advantage to be obtained if a solar energy collector can be located in space because the annual energy accumulation of a sun-oriented area in Earth orbit would be greater by a factor of six than the most favourable solar collection areas on the planet itself. Also, if the orbit is properly selected, solar energy delivery to a satellite can be practically continuous. It requires only suitable transmission systems to ensure that most of the inhabited parts of the Earth derive energy if the orbital vantage point is correctly adjusted. Undoubtedly, costs would be much reduced by the employment of such a technology.

Satellites could be situated in circular orbits in the equatorial plane of the planet at such an altitude that the orbital period coincides with the length of the day, i.e. such that the system remains over a fixed point on the surface of the Earth. There is nothing difficult about this because the approach is now in use as regards communication satellites. As there is a 23·45 ° inclination of the equatorial orbit plane to the ecliptic, the satellite would pass south of the Earth's shadow for 6 months and above it for 6 months. On and near the equinoxes, the satellite would be briefly eclipsed (maximal duration of 70 min). Consequently, the satellite would be illuminated for 99·26 % of every year. In such an orbit, the total direct solar insolation (exposure to the sun's rays) is 1·395 W/m^2.

Perhaps the optimal orbiting mechanism would be a simple mirror providing sunlight to a terrestrially based solar power plant. Since the sun has an optically visible diameter, i.e. it is a disc and not a point, the smallest image which could be produced on the planetary surface by such a geosynchronous orbiting mirror is about 330 km in diameter and therefore, if a 'one sun equivalent' image strength is necessary, the orbital mirror must also be 330 km in diameter. In practice, it would have to be more because of the fact that it is impossible to construct a mirror with 100% efficiency. Other factors may militate against maximal efficiency for such a system, notably the occurrence of clouds.

14.3.1. Microwave Energy Reflectors

The proposal has been made that a passive reflector might be used as a relay between a microwave transmitter located near and energised by a terrestrial electrical power plant and a receiver located near the point of power utilisation. Power is transmitted by the microwave beam. The satellite required here would be much less complex than that of the mirror type alluded to above and the major advantage of the system would be that cloud, rain, etc., would not interfere with transmission of the microwave beam. Also, power-generating elements are on Earth.

14.3.2. Orbital Power Generation

In this system, orbital plants which use solar cells for energy conversion have been suggested. Transmission of the generated power to the Earth would be accomplished by means of microwave beams.

Thermal engines might be utilised, perhaps some sort of faceted reflector which concentrates solar energy received into a cavity absorber. A working fluid, say an inert gas, transfers this energy to a thermal engine which in turn rotates an electric generator. This approach is rather like that of the solar tower energy collector on Earth, except that it is in space.

There is one important matter worthy of mention here also and that is that if advanced design nuclear-power plants become politically unacceptable on the planet, they could possibly be operated in geosynchronous orbit.

14.3.3. Microwave Energy Transfer

Tesla first investigated microwave energy transfer. In an orbital power station, the plant's electrical output could be converted to microwaves later formed into a beam by an antenna and passed through the 35 000-km space gap and the atmosphere, thereafter being aimed to a receiving station and converted to d.c. or a.c. power.

14.3.4. Laser Energy Transmission

Laser systems might be suitable for the space–Earth power link. At the moment, it is only feasible to impinge a laser beam on to a thermal absorber acting as a heating unit for a thermal engine. The efficiency would be low (about 35 %) and, as recent disputes about Christmas lasers in Oxford Street, London, in the winter of 1978 showed, the method is potentially dangerous. Also, cloud could block transmission.

14.4. WIND POWER

Power produced by winds is proportional to the cube of wind velocity. Of course, it has been used for centuries by windmills (cf. Don Quixote!), but nowadays wind-turbine generators and aerogenerators are utilised.

Aerogenerators can be divided into small, intermediate and large types, according to rated capacity (0–9 kW, 10–99 kW and 100 kW–3 MW, respectively). The small units are employed in homes and on farms in several countries and they are particularly useful in the isolated areas where some people live and work. Power based on winds is simply enormous. It has been estimated that the total energy of the atmosphere is approximately 10^{14} MW! Also, since the power generated by winds is proportional to the cube of the wind speed, it is very important to local areas enjoying consistently high winds. A 22 mile/h wind will yield eight times more power than one of half that speed. Suitable wind-power sites have been found in several countries, but great care is necessary in taking wind measurements, especially in rugged terrains. Misinterpretation of wind-power potential has caused failure of wind-power systems in the past. Some data regarding large aerogenerators are given below.

1. USA: The largest wind turbine ever constructed was built near Rutland, Vermont in the 1940s. The tower was 110 ft high and there was a two-bladed rotor 175 ft in diameter and an a.c. generator rated at 1·25 MW at 30 mile/h. The generator was located upwind from the tower and the blades downwind from it, the entire unit being kept oriented into the wind by a wind vane aloft which actuated servomechanisms. Tests were periodically conducted from 1941 to 1945 and routine operation for the Central Vermont Public Service Corporation ended in March 1945 when a blade was lost and, owing to wartime shortages, could not be replaced. In 1974, NASA began constructing an experimental wind-turbine generator for installation at the Lewis Plum Brook Test area, Sandusky, Ohio and this has a 100-ft-high tower with a 125-ft-diameter rotor and a planned output of 180 000 kWh/year as a 460-V, three-phase, 60-Hz alternating current.

2. *UK:* In the middle 1950s, an aerogenerator utilising a pneumatic transmission system was built for the Enfield Cables organisation. The tower was 100 ft high and the blades were 80 ft in diameter. The rated capacity was 0·1 MW at 30 mile/h and the generator unit operated for several years in Algeria.

3. *Denmark:* In the second world war, no less than 13 aerogenerators were constructed in the 70–90 kW range and supplied direct current to the d.c. power grids. For example, at Gedser, an aerogenerator was constructed and this was designed to yield 0·2 MW at a speed of 34 mile/h.

4. *France:* An aerogenerator was constructed at Bourget as early as 1929 and this had a 66-ft-high tower with blades 66 ft in diameter and the d.c. generator was rated at 0·015 MW at 13·4 mile/h.

5. *USSR:* An aerogenerator was installed at Yalta in 1931.

6. *Federal Republic of Germany:* Several aerogenerators were built in the early 1960s with capacities up to 1 MW.

Obviously, in siting such equipment as discussed above, some criteria must be satisfied and these are as follows:

1. the site must enjoy a high normal wind speed;
2. there should be no obstructions for at least a couple of miles upwind;
3. optimally, the site should comprise the summit of a smooth and rounded hill with gentle slopes;
4. a shoreline may constitute a suitable location;
5. a mountain gap can be excellent because it can act as a wind funnel.

Appropriate ecological characteristics of such sites may include growth of trees only to bush size and visual indicators such as downwind streaming of tree branches.

Where the object of the exercise is to produce a sizeable amount of electrical energy and not just the quantity, say, to run a single farm, assemblages of wind turbines will be necessary.

A problem encountered with most solar energy systems and found also in regard to winds is that of energy storage. One possibility is pump storage of water. Another is to utilise the wind to compress air and store it in caves or perhaps in aquifers. Additionally, the generated electricity can be used to produce hydrogen and maybe other fuels.

14.5. PHOTOSYNTHESIS

The chemistry of photosynthesis is well known, hence it is only necessary to point out that this may be summarised thus:

$$6CO_2 + 6H_2O = C_6H_{12}O_6 + 6O_2$$

i.e. atmospheric carbon dioxide and water in the presence of sunlight are used by green plants in order to manufacture carbohydrates, light energy being absorbed by chloroplasts containing chlorophylls. The transformation of this energy into chemical energy is effected by a not fully understood process involving the photolysis (decomposition) of water and the activation of adenosine triphosphate (ATP). This energy-rich ATP later energises the fixation of the CO_2 after a set of reactions so that sugar molecules are formed.

The efficiency of utilisation of the total incident light energy on the planet is naturally very low because of the spacing of plants on land and in the sea, the spacing of chloroplasts in the cells and other factors. In fact, of the 73×10^8 kcal falling on an acre of land annually in the temperate regions, a mere $0 \cdot 1 – 0 \cdot 5 \%$ is fixed by organic material! At best, the efficiency probably only rises to 2%. Despite this, however, it is calculated that the energy converted by the process every year is approximately 100 times greater than the heat of combustion of all the coal mined on the planet annually and no less than 10 000 times greater than *all* the energy derived from water power in any single year!

14.6. RECENT DEVELOPMENTS IN SOLAR POWER UTILISATION

The *New York Herald Tribune* of 7 December 1978 cited some exciting work in Israel.[2] Yehuda Bronicki, President of Ormat Turbines Ltd, Yavne, Israel, employs a turbine spinning from the sun's heat to produce electricity which powers lights illuminating the company's research yard containing a waist-deep water tank the size of a tennis court, i.e. a solar pond which seems likely to constitute a breakthrough in using the sun as a commercially viable source of electrical power. In fact, this is said to be the only turbine now available as a solar power converter and the prediction is made that it will be marketed, together with the solar pond, in another year and a half. The Ministry of Energy believes that Israel and this company in particular hold a considerable lead over the rest of the world in this aspect of market readiness and the turbine used is a normal one which the same company has employed in 2000 electricity-generating units using power provided by fossil fuels. The solar pond or sunpool comprises a salt-water pond covered by a layer of fresh water. As the salt water cannot rise, it retains its heat. A natural one exists on the shore of the Gulf of Aqaba, but, of course, artificial ones are easy to construct.

The turbine generator in question is a self-contained, maintenance free, closed cycle vapour type with only one moving part, the shaft itself. A sample unit was built by Ormat in Mali in 1966 and cost only $US25 000, producing power equivalent to that of a 1-hp pump, but, due to lack of interest on the part of the Mali government, this unit is now back in Israel. Its disadvantage is that it needed a considerable amount of plumbing to carry hot water from 30 solar collectors to the turbine. The solar pond does not require such piping and the minimal size envisaged is about a quarter of an acre. The Ormat unit produces 5 kW of power per hour, enough to supply a small village. Actually, Ormat has also completed a 2-acre solar pond at Sodom on the Dead Sea which will be generating power in the spring of 1979 and is expected to provide hot water and refrigeration for a 200-room hotel under construction in the vicinity.

Ironically, Herblock has a cartoon in the same issue of the *New York Herald Tribune* and this shows the oil companies and the US administration busily tending boilers of natural gas, coal, oil and nuclear power while neglecting one devoted to solar energy development and heated by a single flickering candle!

14.7.   THE GEOTHERMAL RESOURCE

Clearly, solar energy is a very valuable prospect potentially, but its development has not been as rapid as might have been hoped and, to some extent, this is also true of the geothermal resource. Thus, it is important that extensive research into both should continue and, as regards the latter, it is essential that the general public should be made aware that geological sources are a long lived and enormous subterranean reservoir of potential power which is much less affected by withdrawal of heat than fossil-fuel sources are by their exploitation. Besides this, other factors favour their increasing development and utilisation. Among these may be mentioned the fact that seismic activity and subsidence effects do not increase critically when geothermal fluids are removed; also, the water table is not affected sufficiently to harm society or its environment. Apropos the environment, air and water pollution from the impurities in geothermal steam appears to be minimal. There is an impact due to the construction of wells, pipelines and power plants, but this is mostly evident during the development stage and these installations need not necessarily interfere seriously with agricultural activities in the same area. It is true that drilling operations do contribute much noise in the relevant region, but geothermal sources are usually remote and in any case, once the field is in production, the noise level drops. Additionally, mufflers can be employed as at The Geysers during drilling and testing; they lower the noise level significantly. Reinjection of wells with waste water practically obviates the risk of subsidence. Although noxious gases do result from drilling of geothermal wells, the fact that geothermal plants operate without combustion means that they are of a different kind and occur in much smaller quantities than the output from a fossil-fuel plant. Non-condensable gases found in geothermal steam include CO_2, CH_4, H_2, N_2, NH_3 and H_2S; the last, although it is released in lesser amounts than the others, is the most unpleasant from an environmental point of view. It is soluble in water and escapes into the air by evaporation. As regards local waters, they may become degraded because of the dissipation of reject heat in them. However, thermal pollution of this type can be avoided by rejecting waste heat directly into the atmosphere. In the case of dry-steam fields from which waste heat is produced, this approach has been used and hence thermal pollution has been obviated.

Perhaps the most telling point in utilising the geothermal resource to its maximal capacity, however, is that cost analyses demonstrate that producing geothermal power is highly competitive as compared with conventional forms of electrical power production (Fig. 14.1). Nevertheless, it may well prove too expensive a proposition for existing electricity utility companies to handle the financial burden involved in developing it. Certainly the oil industry could do so and has manifested interest within recent years. Several companies such as Standard Oil of New Jersey and Shell have initiated inquiries regarding the matter of geothermal land leasing. As a result, some questions have been raised as to the propriety of permitting future geothermal power development in the USA to pass into private hands and the ownership of the oil industry.[3]

Statistics show that the fixed charges of a geothermal power plant are about the same as those for conventional fossil-fuel plants, but less than

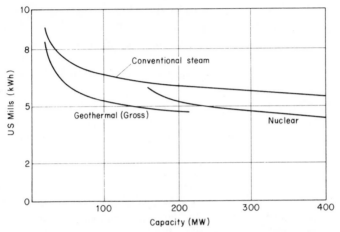

FIG. 14.1.   Estimated production costs of nuclear and conventional thermal plants
compared with tentative geothermal power costs.

half of those for a nuclear or hydropower plant. But, and it is a big but, it
must be remembered that there is a practically constant upward spiral in
conventional fuel costs. In the case of oil, dependence on overseas suppliers
to the USA is a very negative factor which does not apply to geothermal
power. P. Mathews gave some relevant data regarding the average of
geothermal costs between 1961 and 1972 at The Geysers[4] and these were
cited in Chapter 6 (Table 6.4) to which reference may be made. They showed
that there ought not to be much variation because the only significant
variable is the input of steam. Apart from pay increases for personnel, of
whom there are relatively few, other costs are practically constant, even
taking inflation into account.

As indicated earlier, the consistency of geothermal costs depends upon
two major factors, namely

1.   adequate profit,
2.   increasing efficiency, i.e. using all the aspects of a particular
      geothermal resource.

The second parameter involves imponderables outside human control such
as geological characteristics which may favour or impede a geothermal
plant's output of energy and thereby react so as to decrease or increase the
price of this. Consequently, the importance of proper initial feasibility
studies and scientific surveys cannot be over-emphasised.

As regards the actual designing of the power plant itself, perhaps the most vital single matter is correctly to assess the availability of steam. To effect this, the following features should be carefully examined:

1. the type and location of the geothermal field and of high-density steam sites within it;
2. the amounts of steam available;
3. properties such as temperature, pressure and available enthalpy;
4. the quantities of impurities and non-condensable gases present in the steam as well as their nature.

Obviously, the fluid amount in all geothermal prospects is important and so is its quality, i.e. whether it is dry steam, wet steam or hot water. Endogenous fluid does show great variability and each prospect requires individual treatment. None the less, all geothermal fields possess heat, heat produced in many forms and quantities, and this is the source of derived power, that power alluded to by Job (28: 5) thus: 'as for the Earth, out of it comes bread; and under it is turned up as it were fire'.

The fires beneath have a great contribution to make in solving Man's mainly self-created energy problems and the geothermal resource they provide is already locally significant in some places. It is to be hoped that this currently restricted utilisation will expand greatly during the rest of the century and thus make an ever-increasing contribution towards both the alleviation of energy shortages and the reduction of pollution.

REFERENCES

1. Claude, G. (1944). Power from tropical seas. *Mech. Engineering*, **52**, 1039–44.
2. Torgerson, Dial (1978). 'Israeli firm readies cheap solar power turbines'. In: *New York Herald Tribune*, December 7 issue, pp. 1–2.
3. Cheremisinoff, Paul N. and Morresi, Angelo C. (1976). *Geothermal Energy Technology Assessment*. Technomic Publishing Co. Inc., Westport, Conn., 164 pp.
4. Mathews, P. (1973). Geothermal experience at Geysers power plant. *Proc. Amer. Soc. Civil Engrs., J. Power Div.*, **99**, N.PO2, 329–38.

Glossary

The following glossary is intended to define a number of technical words and terms found in the text and covers aspects of chemistry, physics and geology as well as geothermics.

Abyss: That marine environment below 4000 m in which the only sediments are red clay and the deep sea oozes; at this depth, the temperature does not exceed 4 °C.

Aeolian: Pertaining to sediments deposited after transportation by wind.

Aeon: 10^9 years. The age of the Earth is about 4·7 aeons.

Allochthonous: A term used in reference to material forming rocks which has been transported to the site of deposition.

Amphiboles: A group of inosilicates characterised by a double chain of linked SiO_4 tetrahedra. The general formula is $X_{2-3}Y_5Z_8O_{22}(OH)_2$, where X may be Ca, Na or K, Y may be Mg, Fe^{3+} or Fe^{2+}, Al or Ti and Z may be Si or Al (maximum of two Al). The hydroxyl may be partially replaced by F, Cl or O.

Andesite: A fine-grained volcanic igneous rock with low silica content and less than 10 % of quartz together with feldspar, especially.

Anhydrite: An evaporite mineral, $CaSO_4$, found in sedimentary rocks associated with gypsum.

Aquifer: A water-bearing bed or set of strata.

Arenaceous: Pertaining to a group of detrital sedimentary rocks, typically sandstones in which the particles range in size from $\frac{1}{16}$ mm to 2 mm. These accumulate either by wind action or through deposition in water, and in the latter case they may form in marine-water, brackish-water or fresh-water environments. In most arenaceous rocks, the grains are mainly quartz, but feldspars, mica, glauconite and iron oxides may also be present. Cementation may occur and involve calcareous, siliceous or ferruginous cement (*v.* also *Dialysis* below).

Argillaceous: Pertaining to a group of detrital sedimentary rocks, commonly clays, shales, mudstones, siltstones and marls. Two grades of particle size are involved, namely the silt grade with particles ranging in size from $\frac{1}{16}$ mm to $\frac{1}{256}$ mm and the clay grade with particles smaller in size than $\frac{1}{256}$ mm. In addition to clay minerals, argillaceous rocks may contain colloidal material, very finely divided quartz, carbonate dust and finely divided carbon. Argillaceous rocks are almost always deposited in water (fresh, brackish or marine).

Asthenosphere: The lower part of the sima, presumed to have low strength and little rigidity compared with the lithosphere.

Bar: This is the unit of pressure in the cgs system, actually a pressure of 10^6 dyn/cm^2 and equivalent to 0·986 923 atm (about 75 cm of mercury).

Basalt: A fine-grained and occasionally glassy rock of basic igneous type and having low silica content (45–50 %).

Basic rocks: These are quartz-free igneous rocks containing feldspars which are usually more calcic than sodic.

Batholith: Any large intrusive mass of igneous rock (almost invariably granite) with a large outcrop and no observable base.

Benioff zone: At destructive plate boundaries, crust is being consumed by being forced down beneath island arc–ocean trench systems or mountain belts, with the cold lithosphere descending into the asthenosphere as far as 700-km depth. These usually curvilinear features constitute subduction or Benioff zones (after the seismologist Hugo Benioff).

Breccia: One of the two main types of rudaceous rocks. Comprising angular fragments implying minimal transportation before deposition, breccias are poorly sorted. They may be cemented, formed by collapse (in limestone regions), volcanic or fault-originated.

Breeder reactor: A nuclear reactor which produces the same kind of fissile material as it consumes. For instance, a reactor utilising plutonium as a fuel can produce more plutonium than it uses by conversion of uranium-238.

Brine: A heavily saturated salt solution.

British thermal unit (Btu): The quantity of heat necessary to raise the temperature of 1 lb of water through 1°F (equal to 251·997 cal or 1055·06 joules (J)).

Cainozoic: That division of geological time which follows the Mesozoic and ends at the Quaternary. The duration is from 65 million years before the present to 2 million years before the present, i.e. about 63 million years. An alternative spelling of the word is Kainozoic. (For subdivisions of this, see Appendix 1.)

Caldera: A very large crater which may arise by coalescence of several small ones or by repeated explosions or by collapse or by stoping, i.e. the upwelling of magma into overlying rocks from a large underground chamber. Crater Lake, Oregon, USA, is a good example of a collapse caldera. A super-caldera is located in northwest Sumatra, covers over 1800 km^2 and perhaps represents the result of a tremendous mass of intrusive igneous material perforating the crust by explosion and thereafter collapsing.

Calorie (cal): Unit of amount of heat, namely the quantity that is required to raise the temperature of 1 g of water through 1°C.

Cap rock: That impermeable or low-permeability set of strata overlying a geothermal reservoir.

Cataclasis: The process of mechanical fracturation or break-up of rocks usually associated with metamorphism or faulting. It is a term used for both small-scale and large-scale phenomena.

Cataclastic rock: The same as a clastic rock, namely one that has undergone the process of mechanical breakage as opposed to plastic deformation or flowage. Some examples are fault breccias or crush breccias.

Catastrophism: A now discarded hypothesis stating that changes in the Earth occur as a consequence of large-scale and isolated catastrophes of short duration (as opposed to the idea that small changes are occurring more or less continuously, i.e. uniformitarianism).

Clastic: Rocks so described are composed of fragments of pre-existing rocks which have been produced by the processes of weathering and erosion and usually transported to a place of deposition.

Conductivity, thermal (heat conductivity): The rate of transfer of heat along a body by conduction. It is measured in calories flowing per second across a centimetre cube of a substance having a temperature difference of 1 °C on opposite faces (in the cgs system).

Conglomerate: Pertaining to rudaceous rocks consisting of rounded or sub-rounded fragments implying more transportation than breccias do.

Continental drift: The hypothesis which proposes that the existing continents were formed by the break-up of a single land mass and drifted into their present positions.

Continental shelf and slope: The continental shelf is that part of the sea floor adjoining a land mass over which the maximal depth of sea water is 200 m. The outer margin is marked by the continental slope which extends to the abyssal region. Continental shelves constitute portions of continental masses which are locally submerged. The edge of the shelf is regarded as the edge of the continental sialic mass.

Core, the Earth's: This is that portion of the planet below the Gutenberg discontinuity, i.e. from a depth of 2900 km to the centre of the Earth. There is believed to be an outer core which is liquid and an inner core (starting at 5000-km depth) which may be solid with a density perhaps as high as 14·5 g/cm^3 and a composition of nickel and iron. The core temperature is estimated as over 2700 °C with a pressure of 3·5 million bar.

Cristobalite: A crystalline polymorph of silica formed by the inversion of tridymite at 1470 °C.

Crust, the Earth's: That portion of the Earth lying above the Mohorovičić discontinuity and divisible into sialic and simatic layers. Three layers may be distinguished, namely
1. top: 1 km thick and composed of sedimentary rocks;
2. intermediate: about 1·7 km thick, composed either of consolidated sedimentary materials or some modification of the bottom layer;
3. bottom: about 5 km thick and comprising basic igneous material.

Curie (Ci): A measure of the activity of a radioactive substance which is now defined as that amount of a radioactive isotope which decays at the rate of $3·7 \times 10^{10}$ disintegrations per second.

Decay constant: The probability of decay of an atomic nucleus per unit time which characterises a radioactive isotope. It determines the exponential decrease with time t of the activity A thus:

$$A = A_0 \exp(-\lambda t)$$

where A_0 is the initial activity as time $t = 0$ and λ is the decay constant.

Delta: The isotopic composition of a water sample is expressed in terms of the per thousand deviation of the isotope ratio from that of a standard (usually SMOW, standard mean ocean water) and the data are given as delta units defined by

$$\delta = \frac{R - R_{SMOW}}{R_{SMOW}} \times 10^3\%_0$$

where R refers to the isotope ratios D/H or $^{18}O/^{16}O$.

Detrital: A term applied to any particles of minerals or rocks which have been derived from pre-existing rock by the processes of weathering and/or erosion.

Diagenesis: Processes which affect a sediment while it is at or near to the surface of the Earth, i.e. at low temperature and pressure. For instance, loose sediment may be consolidated into a rock by a diagenetic process called lithifaction.

Dialysis: The separation of colloids in solution from other dissolved substances by selective diffusion through a semi-permeable membrane.

Diapir: An intrusion which domes up the overlying cover after piercing lower layers. Salt domes constitute diapiric structures.

Diastrophism: The large-scale deformation of the crust of the earth which produces continents, ocean basins, mountain ranges, etc.

Diatomaceous earth: This is a siliceous sediment comprising 'skeletal' remains of microscopic plants (diatoms); it is very fine grained and absorbent.

Dip: The angle made by a plane or bed (stratum) with the horizontal.

Disintegration constant: Decay constant.

Dolerite: A medium-grained basic hypabyssal rock, mineralogically and chemically identical with a basalt.

Dolomite: As a rock, one containing more than 15 % $MgCO_3$. As a mineral, $CaMg(CO_3)_2$.

Dry steam: Steam with a very low vapour content.

Dyke: A sheet-like body of igneous rock which is discordant, i.e. transgresses bedding planes or structural planes of the host rock.

Dyne (dyn): cgs system unit of force. The force which, acting upon a mass of 1 g, will impart to it an acceleration of 1 cm/s^2. 1 dyn = 10^{-5} newton (N).

Eclogite: A metamorphic rock with a composition similar to that of a basic igneous rock but which has either crystallised or recrystallised under high-temperature and high-pressure conditions. Normal constituents include garnet, pyroxene and sometimes amphibole.

Endogenous: Pertaining to processes and materials originating within the Earth.

Enthalpy: Heat content, *H*. A thermodynamic property of a substance given by

$$H = U + pV$$

where *U* is the internal energy, *p* is the pressure and *V* is the volume.

Environmental isotopes: Those isotopes of which the natural abundance variations may be utilised in hydrological studies. They include the heavy stable isotopes of water, namely deuterium and oxygen-18 (which comprise part of the water molecule itself), as well as radioisotopes such as tritium (which can also enter the water molecule), carbon-14, silicon-32 and members of the uranium and thorium families.

Epeirogenic: Pertaining to movements of generally even character producing small-scale faulting, tilting and warping effects in rocks and produced by adjustments in continental or sub-continental land masses as a result of uplift or depression.

Erg: A cgs system unit of work or energy. The work done by a force of 1 dyn acting through a distance of 1 cm. 1 erg $= 10^{-7}$ joule (J).

Erosion: Wearing away of the land surface by the mechanical action of transported debris, a universal process on Earth.

Eustatic: Pertaining to absolute changes in sea level of world-wide significance.

Evaporite: A sediment resulting from the evaporation of saline water. Most evaporite beds derive from bodies of sea water, but borate and sodium carbonate minerals may originate in lakes. It is suggested that a sabkha environment is involved.

Extrusive: Pertaining to igneous rock which has flowed out on the Earth's surface.

Facies: The sum total of features characterising a sediment as having been deposited in a given environment, these including mineral and fossil contents as well as bedding and structure. Facies characterised by rock types are termed lithofacies and those characterised by their included fauna are termed biofacies. A metamorphic facies is an assemblage of metamorphic rocks considered to have been formed under similar dynamothermal conditions.

Fault: A rock fracture along which an observable amount of displacement has occurred.

Feldspars: The most important single group of rock-forming silicate minerals, basically three-dimensional framework structures in which all four of the oxygen atoms of any silica tetrahedron are shared with adjoining tetrahedra. Na and Ca ones constitute plagioclases, alkali feldspars comprising sanidine, orthoclase, microcline, etc.

Ferromagnesian: Pertaining to rock-forming silicates containing essential Fe and/or Mg.

Fission, nuclear: A nuclear reaction in which a heavy atomic nucleus splits into two roughly equal parts, simultaneously emitting neutrons and releasing large amounts of energy. The process may be spontaneous or caused by the impact of a neutron or an energetic charged particle or a photon.

Fissure: An elongate, narrow and deep crack in the Earth's crust.

Fluid: A substance taking the shape of the vessel containing it. A liquid or a gas.

Flysch: Sediments associated with the Alps and produced by erosion of uprising and developing fold structures, subsequently deformed by later stages in the development of these same fold structures. In Switzerland, such sediments are marine and argillaceous to rudaceous in nature.

Fold: A flexure in rocks.

Foreland: The area of a fold which is inclined in the direction of the thrusting force (*v.* Fig. G.1).

Formation: A number of beds or strata.

Fossil fuel: Any of the natural fuels which are obtained from the Earth and including coal, petroleum and natural gas.

Fold

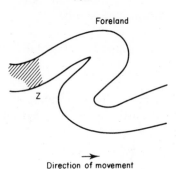

Foreland

Z

Direction of movement

FIG. G.1. Relationship between folding and the development of a geothermal area. Z = zone of pressure relief; shaded area = geothermal region characterised by high heat flow.

Fracture: Any break in a material. In geology, breakage in rocks or minerals not along the direction or directions of cleavage or fissility.

Fumarole: A hot spring emitting volatiles or a vent in a volcanic area from which smoke and gases arise.

Fusion, nuclear: A nuclear reaction between light atomic nuclei as a result of which a heavier nucleus is formed and a large amount of nuclear energy is released.

Gabbro: A coarse-grained plutonic basic igneous rock consisting of basic plagioclase, a pyroxene and usually olivine. Gabbros are low in silica and high in Mg and Ca. Na and K are also low.

Geosyncline: A major structural and sedimentational unit of the Earth's crust comprising an elongated basin which becomes in-filled with masses of sediment and therefore subsides. Volcanic rocks may occur intercalated with sediments. The accumulated mass of sediments is strongly deformed later by orogenic forces and forced into a fold-mountain chain. During this process, the basal portions of the sedimentary mass may be highly metamorphosed and granitic emplacement may also occur.

Geyser: A spring from which columns of boiling water and steam are emitted into the air from time to time.

Glassy: Those parts of rocks not consisting of discrete crystalline units are described as glassy.

Glauconite: Possibly a mica, this is a common constituent of marine sediments of a characteristically bright green colour.

Gneiss: Banded rock formed during high-grade regional metamorphism.

Gondwanaland: The name applied to the hypothetical southern hemisphere supercontinent comprising South America, Africa, Madagascar, India, Arabia, Malaysia and the East Indies, New Guinea, Australia and Antarctica, prior to its break-up under those forces causing continental drift. It corresponds to Laurasia in the northern hemisphere.

Graben: A downthrown block between two parallel faults.

Grain size: In dealing with sediments, precise dimensions must be stipulated for sands, clays, etc., and the ones commonly employed are as follows:

Size range	Particle
Greater than 256 mm	Boulder
64–256 mm	Cobble
4–64 mm	Pebble
2–4 mm	Gravel
$\frac{1}{16}$–2 mm	Sand
$\frac{1}{256}$–$\frac{1}{16}$ mm	Silt
Less than $\frac{1}{256}$ mm	Clay

Granite: A coarse-grained igneous rock comprising essential quartz, alkali feldspar and very commonly mica, biotite and/or muscovite. The accessory minerals may include apatite, zircon and magnetite.

Graywacke: Fine to coarse, angular to sub-angular particles which are mainly rock fragments, usually poorly sorted and even sometimes pebbly locally. The cementing material is usually argillaceous.

Gutenberg discontinuity: A seismically defined discontinuity separating the Earth's mantle from its core and lying about 2900 km below the surface.

Half-life: The time taken for the activity of a radioactive isotope to decay to half its original value, i.e. for half of the atoms present to disintegrate.

Heat: Energy possessed by a substance in the form of kinetic energy of atomic or molecular translation, rotation or vibration. The heat contained by a body is the product of its mass, temperature and specific heat capacity. It may be expressed as joules (SI units), calories or British thermal units.

Hot spot: A location at which a geothermal resource is situated.

Hot spring: A spring created when the amount of subterranean water is too great to be turned into steam by the magma, the heated water flowing on until emerging from holes in the ground.

Hot-water system: A type of geothermal resource existing when water temperatures are usually below the boiling point at atmospheric pressure and they range from 50 °C to 125 °C. The water from such systems is similar to the ground and surface waters of surrounding regions and normally appears as hot springs.

Hydrothermal processes: These are those associated with igneous activity and involve heated or superheated water. Water at very high temperature becomes very active and is quite capable of breaking down silicates and dissolving substances normally considered to be insoluble. There are two major types of hydrothermal activity, namely
1. alteration processes such as kaolinisation and serpentinisation,
2. deposition processes responsible for ores of copper, lead and zinc.

Hypabyssal: Pertaining to a category of rocks, namely intrusive igneous ones which have crystallised under conditions intermediate between plutonic and volcanic. Of medium grain, hypabyssal rocks include dykes and sills. They form nearer the surface than plutonic rocks.

Hyperthermal (geothermal) fields: Comprise wet fields producing pressurised water at a temperature in excess of boiling point. In these, when fluid arrives at the surface, most remains as boiling water, but a fraction flashes into steam. Also comprise dry fields which produce dry, saturated or slightly superheated steam at pressures exceeding atmospheric. Both cases are synonymous with a hydrothermal convective, liquid-dominated system as described in Chapter 2. Hyperthermal fields were so designated by H. Christopher and H. Armstead, 1978, *Geothermal Energy*, E. & F. N. Spon, London.

Igneous rocks: One of three major groups of rocks regarded as the primary source of the material comprising the surface of the Earth. Mostly, they appear to be rocks which have crystallised from a silicate melt, i.e. magma, and may be extrusive (volcanics) or intrusive (hypabyssals and plutonics) in occurrence. They are classified on the basis of silica content, colour, feldspar character, grain size and texture.

Ignimbrite: Welded tuffs. They result from deposition by nuées ardentes, incandescent clouds of gas and volcanic ash violently emitted during certain volcanic eruptions. High temperatures are involved and the materials comprise pyroclastics such as tuff, pumice, lapilli, etc.

Induration: The process whereby soft sediment is transformed into hard rock.

Intrusive: Pertaining to a mass of igneous rock which has forced itself into pre-existing rocks either along some definite structural feature or by deformation and cross-cutting of the invaded rocks.

Island arc: An arcuate chain of islands such as Japan. They lie around the Pacific margin and in the Caribbean, they are associated with strong seismicity and deep-focus earthquakes as well as deep oceanic trenches. The island arc is a region of intense gravitational and magnetic anomalies.

Isostasy (v. Fig. G.2): The tendency of the Earth's crust to maintain a state of near-equilibrium, i.e. should anything occur to modify the existing state, a compensating change will take place to maintain a balance. The best example is the case of deep roots of mountain chains. As erosion occurs and reduces the height of the mountains, compensation in the form of renewed uplift takes place. Also, continental blocks may be

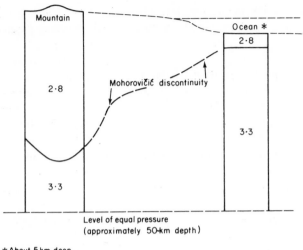

FIG. G.2.   Isostasy. Isostatic balance between a mountain range and the ocean. Below a certain depth in the Earth, the hydrostatic pressure does not vary horizontally. As a result of the fact that mountains are made of low-density material, they float at a higher elevation on the planetary surface than do oceanic basins which are composed of higher-density material. The low-density material (sial) and the higher-density material (sima) do not affect the existence of a level of approximately equal pressure at a depth of about 50 km. Mostly, mountains stand no higher than 6 km, the maximal elevation at which they can maintain isostatic equilibrium. The Himalayas are an exception because they were formed in a continental collision and may be floating upon the combined buoyancy of India and Asia.

depressed under the local load of an ice sheet, recovering isostatically as the ice retreats.

Isotopes: Atoms of the same element, i.e. having the same atomic number, which differ in mass number.

Joint: A fracture in a rock between the sides of which there is no observable relative movement (a distinction from faults).

Joule (*J*): The derived SI unit of work or energy. The work done when the point of application of a force of 1 newton (N) is displaced through a distance of 1 m in the direction of the force. Also, the work done per second by a current of 1 ampere (A) flowing through a resistance of 1 ohm (Ω).

Juvenile water: Water arising from a subterranean magmatic source.

Karroo: A stratigraphic term applied to a series of sediments and lava flows of great extent in southern and central Africa which are Upper Carboniferous to Lower Jurassic in age and contain reptiles and mammal-like reptiles. They include tillites, evaporites, coal seams and red beds.

Kimberlite: Blue ground. A brecciated peridotite containing essential mica, bronzite and chrome-diopside. Diamonds occur.

Kraton: A synonym for shield, also spelled craton.

Land bridge: A hypothetical transoceanic connection between continents which permitted the migration of terrestrial organisms. Since the development of plate tectonics, the idea has been abandoned.

Laser: Light amplification by stimulated emission of radiation. An optical maser (microwave amplification by stimulated emission of radiation). The laser produces a powerful and highly directional monochromatic and coherent beam of light and works on the same principle as the maser except that the active medium either comprises or is contained in an optically transparent cylinder with a reflecting surface at one end and a partially reflecting surface at the other. Stimulated waves make repeated passages up and down the cylinder, some of them emerging as light through the partially reflecting end.

Laurasia: The name given to the hypothetical northern hemisphere supercontinent of North America, Europe and Asia north of the Himalayas prior to its break-up under the forces causing continental drift. It corresponds to Gondwanaland in the southern hemisphere.

Lithifaction: That process of consolidation which produces a hard rock from a loose sediment.

Lithosphere: The outer and rigid part of the crust of the Earth comprising the surface rocks (sial) and the upper part of the sima. This part of the crust has a high strength compared with the asthenosphere.

Maar: The explosion vent of a volcano.

Magma: A molten fluid formed within the crust or upper mantle of the Earth which may consolidate to form an igneous rock. It comprises a complex system of molten silicates with water and gaseous material in solution.

Magnetite: An iron mineral, Fe_3O_4.

Mantle: That portion of the Earth's interior lying between the Mohorovičić and Gutenberg discontinuities, i.e. from about 35 km down to 2900 km. Densities range from $3 \cdot 3 \, g/cm^3$ at the Mohorovičić discontinuity to $5 \cdot 7 \, g/cm^3$ at the Gutenberg discontinuity. The mantle probably consists of olivine principally.

Mesozoic: An era ranging in time from 230 million years before the present to 70 million years before the present and therefore having a duration of about 160 million years. It was preceded by the Palaeozoic and followed by the Tertiary. (For subdivisions of this, *v.* Appendix 1.)

Metamorphic rocks: These result from metamorphism. Metamorphic processes are considered as occurring in the solid state and their end products, i.e. metamorphic rocks, result from the interaction of the metamorphic agencies with the parent rock. Metamorphic processes are classified as isochemical when no introduction of material from an external source is involved or metasomatic when alteration of the bulk composition of the rock is involved. There are five types of metamorphism, namely
1. thermal, involving heat alone,
2. dynamic, involving intense localised stresses tending to fragment rocks,
3. regional, resulting from large-scale heat and pressure,
4. retrograde, a reversal of high-grade metamorphism to produce rocks of a lower metamorphic grade,
5. autometamorphism, involving changes which occur during the

cooling of an igneous mass resulting from the activity of residual fluids within the mass.

Metamorphism: The processes by which changes are caused in the Earth's crustal rocks by the agencies of heat, pressure and chemically active fluids.

MeV: 10^6 eV (electron volt), 1 eV being the work done on an electron when passing through a potential rise of 1 volt.

Micron: Micrometre (μm), one millionth of a metre (10 000 Ångström units).

Microwaves: Electromagnetic radiation with wavelengths ranging from very short radio waves almost to the infrared region (i.e. wavelengths from 30 cm to 1 mm).

Mofette: A type of fumarole.

Mohorovičić discontinuity: Sometimes called the Moho, this is an outer layer, seismically defined in the Earth separating the crust from the mantle, its *average* depth being 35 km.

Mud volcano: A hot spring in a volcanic area incorporating enough ash or mud to produce a boiling mass.

MW: 10^6 watts.

Nappe: Thin, broad moving sheet of rock which is discontinuous with the rock below, being separated from it by a thrust fault.

Nuée ardente: An incandescent cloud of gas and volcanic ash violently emitted during the eruption of a certain type of volcano.

Obsidian: Pitchstone, a black glassy variety of the acid volcanic igneous rock rhyolite.

Olivine: Group name for a set of rock-forming minerals, silicates, with a general composition of $R_2^{2+} SiO_4$ where R^{2+} may be Mg, Fe^{2+}, Mn or Ca (part). The essential structure comprises a series of isolated SiO_4 tetrahedra which are linked by means of metal cations.

Opal: A hydrated amorphous variety of silica, probably derived from a silica gel. Opal occurs primarily as a secondary deposit formed by the action of percolating ground water.

Ophiolite: Ophiolite complexes are found in most mountain belts and consist of masses of basic or ultrabasic igneous rocks such as basalt, gabbro and peridotite with structure suggesting oceanic crust which has been sheared from descending plates and forced up under compression into overlying rock. They provide important clues to the presence of former subduction zones now enclosed by continent.

Orogeny: This is a period of mountain-building, orogenesis being the name given to the process leading to the formation of the intensely deformed belts constituting mountain ranges.

Osmosis: The flow of water or other solvent through a semi-permeable membrane which permits the passage of the solvent itself, but not of dissolved substances.

Outcrop: The total area over which a particular rock unit occurs at the surface of the crust whether visibly exposed or not.

Palaeomagnetism: This is the study of change in the positions of the Earth's magnetic poles during geological time.

Palaeozoic: The era ranging in time from 600 million years before the present to 230 million years before the present and therefore lasting 370 million years. (For subdivisions of this, *v.* Appendix 1.)

Pangaea: The hypothetical original single super supercontinent supposed by Alfred Wegener to have existed prior to becoming fragmented (about 300 million years ago) by continental drift forces to produce Laurasia and Gondwanaland. The latter became separated by an extensive seaway, the Tethys.

Peridotite: A class of ultra-basic rocks consisting mainly of olivine with or without other ferromagnesian minerals.

Permeability: That property of a rock which enables water or some other liquid to pass from the upper surface to the lower surface through its body and resulting from the fact that the rock is either porous or pervious.

Perviousness: This rock property exists where the rock mass is characterised by mechanical discontinuities such as joints, faults, bedding planes, etc., thus allowing water or some other liquid to pass through it.

Phyllite: A cleaved metamorphic rock having affinities with slates.

Pillow lava: Lava in the form of distorted globular masses which formed under water.

Plate tectonics: Modern studies of the major structural features of the planetary crust suggest that these may be employed so as to define a series of regions of this called plates. The typical plate includes the continental shelf, sea and oceanic areas as well as the continent. The theory of continental drift requires relative movement between continental blocks but in the theory of plate tectonics, the movement is rather between plates. Features such as transcurrent faulting represent the lateral movement of plates, mid-oceanic ridges result from plates separating accompanied by sea-floor spreading and trenches associated with island arcs occur where one plate moves under another, i.e. is subducted. Other features such as orogeny, seismicity and volcanicity can be related to plate movements.

Plutonic: An adjective describing a mass of igneous material presumably of deep-seated origin.

Pneumatolysis: Those changes resulting from the action of hot gaseous substances (not water) associated with igneous activity.

Porosity: A rock is described as porous if it contains voids between the mineral grains which can contain liquid. A porous rock is not necessarily permeable. For instance, clay is porous, i.e. can contain water, but is also impermeable, i.e. will not allow water to pass through it.

Precambrian: That period of time from the consolidation of the crust of the Earth to the base of the Palaeozoic. The duration is probably in excess of 4 aeons. (For subdivisions of this, *v.* Appendix 1.)

Pumice: A highly vesicular material derived from acid lavas and produced in large quantities.

Pyrite: The commonest sulphide mineral, FeS_2.

Pyroclastic rocks: Rocks comprising fragmental volcanic material which has been blown into the atmosphere by volcanic activity. Within this group occur pumice, scoriae, tuffs and ignimbrites.

Pyroxenes: A group of rock-forming silicate minerals having a typical chain structure of linked SiO_4 tetrahedra (inosilicates) giving a unit of the form $(SiO_3)_\infty$. They resemble the amphiboles, but have a cleavage of $90°$ as opposed to $124°$ for the latter.

Pyrrhotite: A mineral, Fe_nS_{n+1}, found in basic igneous rocks and contact metamorphic rocks.

Quartzite: A metamorphosed arenaceous rock, usually a sandstone. Constituent grains recrystallise and develop an interlocked mosaic texture with almost no trace of cementation. Impurities in the original rock will produce metamorphic minerals, for instance chlorite or biotite. This type of rock may result from thermal or regional metamorphism.

Quaternary: The latest time interval in the stratigraphic column, 0 million years before the present to 2 million years before the present. The duration is 2 million years.

Radioactive (radiometric) dating: A dating method based upon the fact that the rate of decay of a radioactive element is a constant. This rate of decay may be expressed as the half-life. Where $T_{1/2}$ is the half-life and λ is the rate of decay, $T_{1/2} = 0·693/\lambda$.

Remanence: The residual magnetisation of a ferromagnetic substance

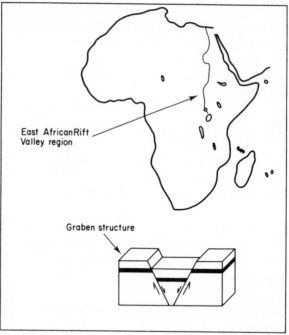

East AfricanRift
Valley region

Graben structure

FIG. G.3.   Rift valley structure. This is a potential geothermal prospect and
investigations have been carried out in East Africa, for instance by Hunting
Geology and Geophysics Limited in Ethiopia and Kenya for the United Nations. In
Ethiopia, 7600 line km of pre-dawn airborne infrared linescan survey were involved.
In Kenya, the same method was applied in two areas in the rift valley.

subjected to a hysteresis cycle when the magnetising field is reduced to
zero.

Rhyolite: Fine-grained to glassy acid volcanic rock.

Rift valley (v. Fig. G.3): An elongated trough bounded by faults. A graben.
   Volcanic activity is characteristically associated with rift valleys. They
   may result from tension created during the fragmentation of continental
   masses or from the slow up-doming of a kraton.

Rudaceous rocks: A group of detrital sedimentary rocks in which the
   particles range in size from 2 mm upwards. They can be deposited in
   water or sub-aerially and comprise two main classes, breccias and
   conglomerates.

Sabkha: Broad, salt-encrusted, supra-tidal coastal flats bordering lagoonal or inner-shelf regions, for instance at Abu Dhabi. Such areas are only occasionally flooded. There is also a continental variety.

Salt domes: Diapiric masses of salt. Salt deforms plastically under high pressure and behaves like an intrusive magma. It can then deform and rupture overlying sediments.

Santa Catharina: A stratigraphic term for the equivalent of the Karroo in Brazil.

Sedimentary rocks: Rocks formed from material derived from pre-existing rocks by processes of denudation together with material of organic origin. The term includes both consolidated and unconsolidated materials.

Semithermal fields: These are geothermal regions which can produce hot water at temperatures up to boiling point from depths of 1–2 km. A specialised category of hydrothermal convective, liquid-dominated geothermal system as described in Chapter 2 and so designated by H. Christopher and H. Armstead, 1978, *Geothermal Energy*, E. & F. N. Spon, London.

Serpentine: A layer lattice mineral, $Mg_6Si_4O_{10}(OH)_8$.

Shield: A major structural unit of the Earth's crust consisting of a large mass of Precambrian rocks (both metamorphic and igneous) which has remained unaffected by later orogenies.

Sial: The upper portion of the Earth's crust composed predominantly of silica and aluminium.

Sima: The lower portion of the Earth's crust composed primarily of silica and magnesium.

SMOW: Standard mean ocean water, an isotopic standard for isotope ratio measurements in water. Based on the original PDB (PeeDee Belemnite) standard of Harold C. Urey and his associates.

Sphene: $CaTiSiO_5$, a mineral found as an accessory in acid igneous rocks and in metamorphosed limestones. It is sometimes called titanite.

Steam: Water, H_2O, in the gaseous state; water above its boiling point. An invisible gas. White clouds emitted by geysers (or kettles) comprise droplets of liquid water condensed from steam.

Stock: An intrusive mass of plutonic igneous rock similar to, but smaller in size than, a batholith.

Stratigraphy: The study of the stratified rocks (sediments and volcanics) and especially their sequence in time, their characteristics and the correlation of beds in different localities.

Strike: The direction in which a horizontal line can be drawn on a plane. In geology, a line drawn at right angles to the true dip.

Subduction zone: Another term for a Benioff zone.

Tectonics: A term pertaining to changes in the structure of the Earth's crust, the forces responsible for such deformations and the external forms produced.

Telluric: Pertaining to the Earth or to soil.

Tertiary: A period of time between the end of the Mesozoic and the Quaternary and lasting about 63·5 million years.

Tethys: A Mesozoic geosyncline which developed between Laurasia and Gondwanaland and covered southern Europe, the Mediterranean, North Africa, Iran and the Himalayas, becoming a seaway before culminating in those mountains and others such as the Alps.

Tillite: Boulder clay, material left by ice.

TNT: Trinitrotoluene, $C_7H_5(NO_2)_3$.

Travertine: A type of calcareous tufa deposited by certain hot springs in volcanic areas.

Trench: An oceanic deep.

Tufa, calcareous: Deposit of $CaCO_3$ formed by deposition from a solution of calcium bicarbonate.

Tuff: Unconsolidated material in ash when it consolidates constitutes this. Ejected from volcanoes.

Uniformitarianism: A fundamental concept in geology, i.e. that processes operating at present also operated in the past and produced the same results. The processes need not have operated at the same rate or at the same intensity as they do today.

Volatiles: Elements and compounds dissolved in a silicate melt and which would be gaseous at the temperatures involved if it were not for the high pressures and solvent effects of the magma. The commonest ones are water and CO_2, but among the rest are Cl_2 and HCl, F_2 and HF, S, SO_2 and H_2S as well as, occasionally, NH_3 and borates. Volatiles have a great effect upon magma because they can
1. lower its viscosity, thus promoting mobility,
2. depress the freezing point, thus allowing it to remain liquid to a much lower temperature and therefore prolonging the period during which crystallisation can take place,
3. react with earlier-formed minerals and thus produce alteration products.
They are particularly important in volcanic rocks and, in geothermal areas, steam is the commonest and most significant volatile.

Volcanism: Volcanic activity. A volcano is a vent in the crust of the Earth through which molten magma, hot gases and fluids escape to the land surface or bottom of the sea.

Watt: The energy expended per second by an unvarying electric current of 1 ampere (A) flowing through a conductor the ends of which are maintained at a potential difference of 1 volt (V). Equivalent to 10^7 ergs/s.

Weathering: The process by which rocks are broken down and decomposed by the action of external agencies such as wind, rain, changes of temperatures and organisms. There are two main types, namely
1. mechanical—mainly caused by temperature change;
2. chemical—mainly caused by the action of substances dissolved in rain water.

Weichert–Gutenberg discontinuity: Another name for the Gutenberg discontinuity.

Welded: A descriptive term used to indicate certain pyroclastic rocks in which the particles have cohered through the heat which they retained and the associated gas immediately after deposition.

Well logging: Techniques of measurement of physical properties in boreholes and including

1. Spontaneous or self-potential (SP) logs—measurement of variations in potential due to the natural currents which flow in the circuit formed by less permeable strata such as shale, the drilling fluid and more permeable strata such as sandstone.
2. Resistivity logs—measurement of the resistivity of strata by means of electric currents applied with a multi-electrode sonde.
3. Gamma-ray logs—measurement of the natural gamma-ray activity of the strata using a scintillometer.
4. Neutron logs—measurement of induced gamma-ray activity due to the capture of neutrons emitted from a suitable source.
5. Acoustic velocity logs—measurement of the transit time of acoustic energy from a source through the strata to a receiver gives the velocity which varies with the relative path lengths through solids and liquids and therefore can be an index of formation porosity.
6. Temperature logs—variation of temperature in a borehole due to differential rates of heat exchange between drilling fluid and formations can provide information regarding the nature of the strata.

Wet steam: Steam possessing a high vapour content.

Zeolites: A group of silicates containing true water of crystallisation; one of the few mineral groups showing reversible dehydration. The cations usually include Ca, K, Na and Ba. Zeolites demonstrate the property of base exchange and sodium zeolites were originally utilised as water softeners.

Zoisite: A silicate mineral belonging to the epidote group and containing both SiO_4 and Si_2O_7 silicon–oxygen units as may be seen from the formula

$$Ca_2Al_3O . Si_2O_7 . SiO_4(OH)$$

Bibliography

For further details on various aspects of geology, the reader is recommended to read the following:
1. On the Earth:
    1. *Understanding the Earth* by Ronald Fraser. Pelican Books, Harmondsworth, 1967.
    2. *A Revolution in the Earth Sciences (From Continental Drift to Plate Tectonics)* by A. Hallam. Clarendon Press, Oxford, 1973.
    3. *Continental Drift (A Study of the Earth's Moving Surface)* by D. H. Tarling and M. P. Tarling. Pelican Books, Harmondsworth, 1972.
2. On general geology:
    1. *Geology Today* by various consultants. CRM Books, Del Mar, California, 1973.
3. On petrology, the study of all aspects of rocks:
    1. *Textbook of Petrology, Vol. 1: Petrology of the Igneous Rocks* by F. H. Hatch, A. K. Wells and M. K. Wells. Thomas Murby & Co., London, 1972.
    2. *Textbook of Petrology, Vol. 2: Petrology of the Sedimentary Rocks* by J. T. Greensmith, F. H. Hatch and R. H. Rastall. Thomas Murby & Co., London, 1972.
4. On mineralogy, the study of minerals:
    1. *Rock-Forming Minerals* by W. A. Deer, R. A. Howie and J. Zussman. Longman, London. Vol. 1, 1962; Vol. 2, 1963; Vol. 3, 1962; Vol. 4, 1963; Vol. 5, 1962.
5. On volcanology:
    1. *Volcanoes as Landscape Forms* by C. A. Cotton. Whitcombe, Wellington, New Zealand, 1944.
    2. *Volcanoes and their Activity* by A. Rittmann, translated from German by E. A. Vincent. Interscience, New York, 1962.

227

6. On geochemistry:
   1. *Principles of Geochemistry* by Brian Mason. John Wiley and Sons Inc., New York, London, Sydney, 1966.
7. On structural geology:
   1. *Elements of Structural Geology* by E. Sherbon Hills. Chapman and Hall Ltd, London, 1965.
8. On water:
   1. *Ground Water Hydrology* by David K. Todd. John Wiley and Sons Inc., New York, London, Sydney, 1959.
   2. *Water in the Service of Man* by H. R. Valentine. Pelican Books, Harmondsworth, 1967.
9. Dictionary of geology:
   1. *The Penguin Dictionary of Geology* by D. G. A. Whitten and J. R. V. Brooks. Penguin Reference Books, Harmondsworth, 1972.

For information regarding nuclear methodology in the earth sciences, two books may be recommended and these are as follows:

1. *Nuclear Techniques for Mineral Exploration and Exploitation.* Proceedings of a Panel, Cracow, 8–12 December 1969, Panel Proceedings Series, International Atomic Energy Agency, Vienna, STI/PUB/279, 1971.
2. *Guidebook on Nuclear Techniques in Hydrology.* Technical Reports Series Number 91, International Atomic Energy Agency, Vienna, 1968.

Valuable general geothermal publications include the following:

1. *Proc. UN Conf. on New Sources of Energy, Rome*, August 1961. Vols. 2 & 3. UN Geneva. E/CONF. 35/3 and E/CONF. 35/4. Document sales numbers 63.1.36 and 63.1.37.
2. *Proc. UN Symp. on Development and Utilisation of Geothermal Resources, Pisa*, September–October 1970. *Geothermics, Sp. Issue,* **1** and **2**.
3. *Proc. 2nd UN Symp. on Development and Use of Geothermal Resources, San Francisco*, May 1975. Vols. 1, 2 & 3. Lawrence Berkeley Laboratory, University of California.
4. *Unesco: Geothermal Energy—A Review of Research and Development.* Earth Sciences 12, Paris.

The Major Divisions of Geological Time

DURATIONS OF PERIODS AND EPOCHS

CAINOZOIC ERA	Quaternary (Pleistocene and Recent epochs)		about 2 million years
	TERTIARY	Pliocene	about 5 million years
		Miocene	19 million years
		Oligocene	12 million years
		Eocene	16 million years
		Palaeocene	11·5 million years
MESOZOIC ERA	CRETACEOUS	Upper (35 million years) Lower (36 million years)	71 million years
	JURASSIC		about 57 million years
	TRIASSIC		about 32 million years
PALAEOZOIC ERA	PERMIAN		55 million years
	CARBONIFEROUS	Upper (45 million years) Lower (20 million years)	65 million years
	DEVONIAN		50 million years
	SILURIAN		about 40 million years
	ORDOVICIAN		about 65 million years
	CAMBRIAN		70 million years
PRECAMBRIAN ERAS			over 4 aeons

The Precambrian eras have sometimes been termed Archaean (basement complex) for the very oldest rock formations and the Proterozoic for the younger. Also, the Cainozoic, Mesozoic and Palaeozoic are sometimes grouped together as the Phanerozoic (containing visible fossils).

The oldest known unmetamorphosed sedimentary rocks are between $3·3 \times 10^9$ and $3·5 \times 10^9$ years old and show quite unmistakably that they are the products of chemical weathering on land and subsequent deposition

in water, i.e. presumably in an ancient ocean. CO_2 must have been an important weathering agent and the hydrologic cycle was well established already. The atmosphere lacked free oxygen and may have possessed methane as its dominant carbon gas at the very earliest stage of planetary history. Hydrogen escape must have led to the replacement of CH_4 by CO_2 at the time these rocks were formed. Animals are believed to have evolved between 0.6×10^9 and 0.8×10^9 years ago, by which time there had to be enough oxygen in the atmosphere to permit normal respiration.

Side-Looking Airborne Radar (SLAR)

The side-looking airborne radar is an extremely valuable observational instrument, a radar geology tool, which can be utilised in order to produce planimetric, geomorphic, geologic and other maps in areas obscured as regards visible or infrared methods by clouds or atmospheric haze. It is equally effective by day and by night, but it does not yield such detailed interpretations as ordinary aerial photography. Nevertheless, it can be used where the latter is not feasible. It is not surprising, therefore, that cost comparisons between the two approaches are not possible because, in practice, they are almost always employed under different conditions and hence not within the same area. The principle of both is the same, however, namely shape, pattern and texture of a region. Such physiographic features as mountains, plains, drainage systems, etc., are easily recognised, but so too are towns, waterways, airfields, etc. In commercial systems, objects as small as a few metres across can be distinguished. The variations in tone found in radar imagery stem from causes different from the factors of brightness and colour determining photography, actually from the reflectivity characteristics of the target surface (in which *attitude* plays an important role). Naturally, as a variety of wavelengths is available to operators of SLAR it must be borne in mind that the one selected is significant as a consequence of the fact that the reflectance of a particular surface varies with that utilised. Additionally, the penetration of the surface increases with wavelengths.

FURTHER READING

Martin-Kaye, P. H. A. and Williams, A. K. (1972). Radargeologic map of Eastern Nicaragua. *IXth Inter-Guyana Geological Conference.*

Nunnaly, N. R. (1969). Integrated landscape analysis with radar imagery. *Remote Sensing of the Environment* **1**(1), 1–6.

Index of Geothermal Localities

Author Index

Subject Index

Explosives
  conventional, 43, 86–9
  nuclear, 43, 89–92

Flysch, 17, 186, 210
Fossil fuels, 1, 2, 27, 43, 82, 134, 170,
  172
Fumaroles, 26, 27, 31, 44, 47–9, 96,
  103, 138, 141, 160–2, 188, 211

Gasbuggy experiment, 95
Gas-dominated fields, 125–7
Geochemical survey, 47–59
Geological survey, 47
Geomagnetic field, 8, 14
Geopressurised systems, 31, 178
Geothermal Energy Research
  Development and Demonstration
  Act 1974, 138
Geothermal gradient, 31, 60, 157
Geothermal Steam Act 1970, 138
Geothermometers, 48–54, 73, 157–9
Geysers, 21, 27, 31, 36, 96, 146, 212
Glacial deposits, 6
Glomar Challenger, 16
Goldfield (Nevada), 16
Gondwanaland, 5–7, 9, 212
Gorda Ridge, 22
Gravimetry, 64, 103, 104, 177
Gravity anomalies, 103
Greenhouse heating, 131, 134, 137,
  142
Gutenberg discontinuity, 12, 213

HCMM satellite, 65
Heat, 3, 22, 27, 33, 34, 36, 41, 43, 44,
  48, 72, 143
  flow, 3, 12, 19, 22, 28–30, 33, 44,
    60, 61, 72, 141
Helium variations in soil gas, 49
Himalayan mountain belt, 18, 22, 215
Hot springs, 21, 24, 36, 37, 44, 47–50,
  53, 54, 77, 96, 138, 162, 188
Hveravellir–Husavik supply conduit,
  132

Hydroelectric power, 2, 26, 167
Hydrological survey, 47
Hydrothermal convective systems, 28,
  30, 32, 40, 57, 72–85, 213
Hyperfiltration, 153
Hyperthermal fields, 33, 213

Ice ages, 6, 19, 20, 190
Imagery enhancement, 65
Infrared surveys, 46, 61
Integrated system to produce
  electricity, 181, 182
Island arcs, 14, 17, 19, 20, 214
Isotopic geothermometry, 55, 56

JOIDES programme, 15, 16, 19

Karroo system, 6, 21, 216
Kimberlites, 6
Kut Block (Iran), 189

Land bridges, 4, 216
Laser energy transmission, 195
Laurasia, 5, 7, 216
Liquid-dominated systems, 30–3, 36,
  40, 45
Lithium concentrations in thermal
  areas, 49
Lithosphere, 16–20, 22, 217
Los Alamos experiment, 179–81

Magma, 20, 22, 28, 31, 36, 39, 75, 96,
  103, 183, 217
Magnetic anomalies, 13, 14, 17, 177
Magnetic cleansing, 10
Magnetisation, 7–9, 13, 14
Mantle of Earth, 4, 11, 12, 14, 18, 20,
  22, 24, 33, 36, 217
Mendocino Escarpment (off
  California), 13
Mercury in soils of geothermal areas,
  49
Microwave radiometry, 64, 195